IMAGES
of America

SCOTTISH RITE CATHEDRAL

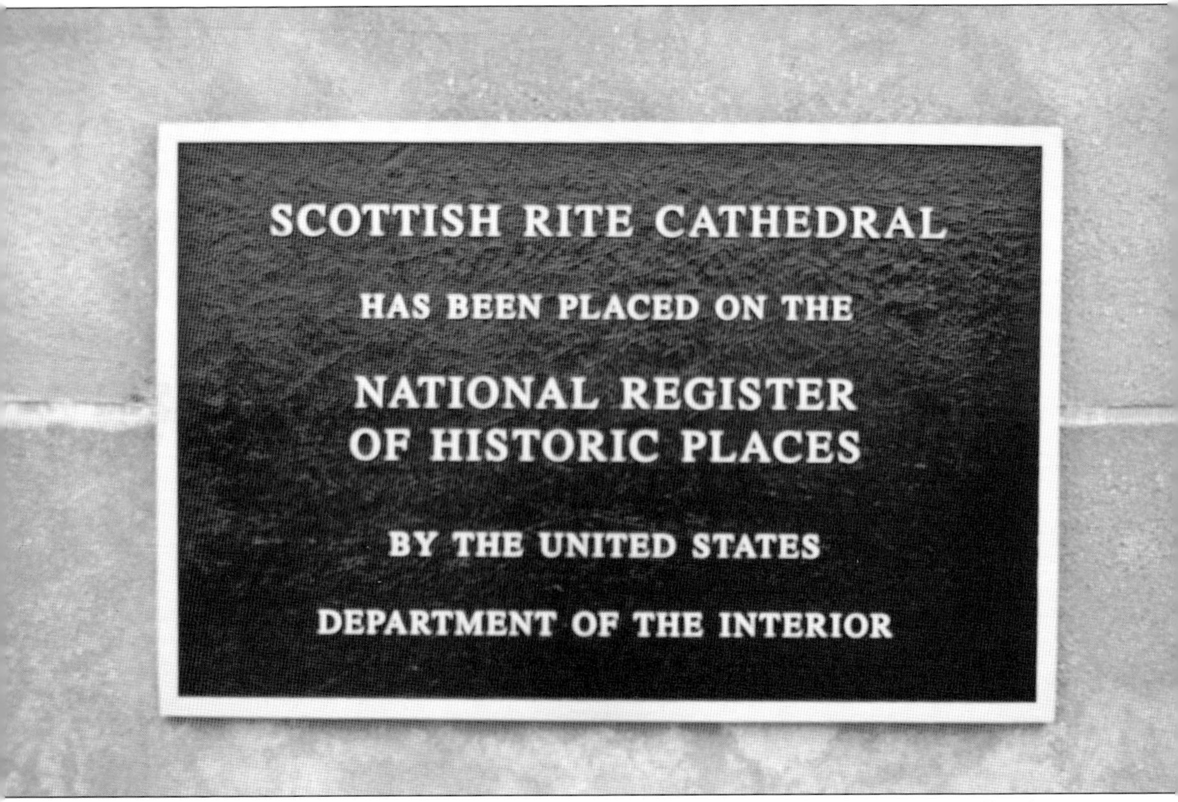

National Register Plaque. This sign on the front of the Cathedral commemorates the placement of the building in the National Register of Historical Places. US representative Jason Altmire and Pennsylvania Historical and Museum Commission's William Callahan delivered speeches at a formal presentation. The world-famous Glenn Miller Orchestra provided entertainment on the Cathedral main stage to close out the evening.

On the Cover: The Scottish Rite Cathedral, constructed in 1925 on New Castle, Pennsylvania's Lincoln Avenue, received a listing in the National Registry of Historical Places in 2009.

IMAGES
of America

SCOTTISH RITE CATHEDRAL

Rob Cummings and Dale Perelman

ARCADIA
PUBLISHING

ISBN 978-1-4671-2733-2

Published by Arcadia Publishing
Charleston, South Carolina

Printed in the United States of America

Library of Congress Control Number: 2017958424

For all general information, please contact Arcadia Publishing:
Telephone 843-853-2070
Fax 843-853-0044
E-mail sales@arcadiapublishing.com
For customer service and orders:
Toll-Free 1-888-313-2665

Visit us on the Internet at www.arcadiapublishing.com

US Postage Stamp. In 2009, the country's postal system issued a 42¢ stamp to celebrate the Department of the Interior's naming of the Scottish Rite Cathedral to the National Register of Historic Places.

CONTENTS

ACKNOWLEDGMENTS

The authors required information from dozens of cathedral experts to bring together the pictures and history of this wonderful building. We especially would like to acknowledge the assistance of Cathedral Foundation executive director Dawn Piroli and the foundation board members, building supervisor Gordon Rees; maintenance worker Brady Robertson; entertainment chairman Jim McKim; stage manager Bob McKibben; past commander of the Valley of New Castle George Morrow; photographer Jim Camp; current commander in chief of the Valley of New Castle Jeff Jackson; foundation president Ken Shiderly; secretary of the Valley of New Castle Randy Connell; and Arcadia editor Stacia Bannerman. The New Castle City Library, the *New Castle News*, and the Lawrence County Historical Society provided background information regarding the ground breaking and construction of the building. Lastly, we would like to give a special thank-you to our technical advisor Bryan Cummings and photographer Bill Conkle, without whom this book would not have been possible.

Unless otherwise noted, all images in this book are courtesy of the Cathedral Foundation.

Introduction

Shortly after the end of World War I, a spirit of optimism surged through Lawrence County. The area's Masons numbered nearly 6,000 dedicated men who met at the Temple Building on North Street in downtown New Castle. The economic and physical growth of the region encouraged a group of Masons to discuss the construction of a larger structure. John Scofield Wallace, the first Commander in Chief of the Valley of New Castle, headed a committee of interested members who scouted the possibility for a more substantial facility to house the increasing membership. Three times the leaders chose a potential building site, each larger than the one before it, but none large enough to hold the present Cathedral.

Lincoln Avenue represented the community's finest real estate and provided a plum view overlooking the downtown area. Based on a consensus of Masonic leadership, the fraternity accumulated property with the idea of constructing a grand cathedral. The purchase of two large homes, formerly owned by the prominent Knox families, provided the bulk of the property required. A committee hired the Milwaukee firms of R.G. Schmidt as architect and S.M. Seisel as general contractor. In 1924, a demolition team razed the two beautiful homes and cleared the property.

Unfortunately, blue clay ran through the land, causing a lack of stability for a foundation. The original spot selected for the building required more work than the group could afford. The second location entailed the installation of 916 concrete pilings to support an adequate footing. The 168,271-square-foot Cathedral was planned to cost around $1 million but actually came in over budget at $1.6 million. Although the architect designed the front to face where the rear of the building sits, financial considerations required a change of plans. After the building committee raised funds through a combination of donations, work in kind, bonds, and loans, construction began in earnest during the fall of 1924.

Although the interior required several more months for completion, Pennsylvania's Right Worshipful Grand Master laid the cornerstone for the Scottish Rite Cathedral in an elaborate ceremony on June 10, 1925. The Fall Reunion, incorporating the initiation of several hundred new Consistory members on November 8, 1926, became the first official use for the entire building.

Today, the Cathedral Foundation Board, a 501(c) tax-exempt organization, manages business and financial matters pertaining to the activities and upkeep of the Scottish Rite Cathedral, which is home for thousands of Masons. Not only is the building an active community venue for entertainment and banquets, but it is also home to the following Masonic bodies and organizations: the Masonic Dyslexia Learning Center of Western Pennsylvania; Mahoning New Castle Lodge 243 and Lodge of the Craft Westminster 433, comprising the first through third degrees; Valley of New Castle Scottish Rite Lodge of Perfection, Princes of Jerusalem, Chapter of Rose Croix, and Consistory; the New Castle York Rite, Christian-based branch of the Masons, which includes the uniformed Lawrence Commandry Knights Templar No. 62, Royal Delta Arch Chapter 170, and Hiram Council No. 45 Royal and Select Master Masons; the Wimodausis, which is made up of the wives, widows, mothers, daughters, and sisters of Master Masons; Legion of Honor, Consistory members comprised of US veterans; High Twelve, whose luncheons occur monthly to discuss Masonic topics of interest; International Order of the Rainbow Girls (IORG), which teaches leadership training through community service to girls between the ages of 11 and 21; and lastly, Shriners International, an organization based on fun, fellowship, and the Masonic principles of brotherly love and truth.

In addition, the Cathedral remains the seat of the Lawrence County Foundation, a philanthropic community funding organization.

In past years, the Eastern Star Chapters 105, 333, and 440, the White Shrine of Jerusalem, DeMolay, the Builders, and other organizations met at the Cathedral.

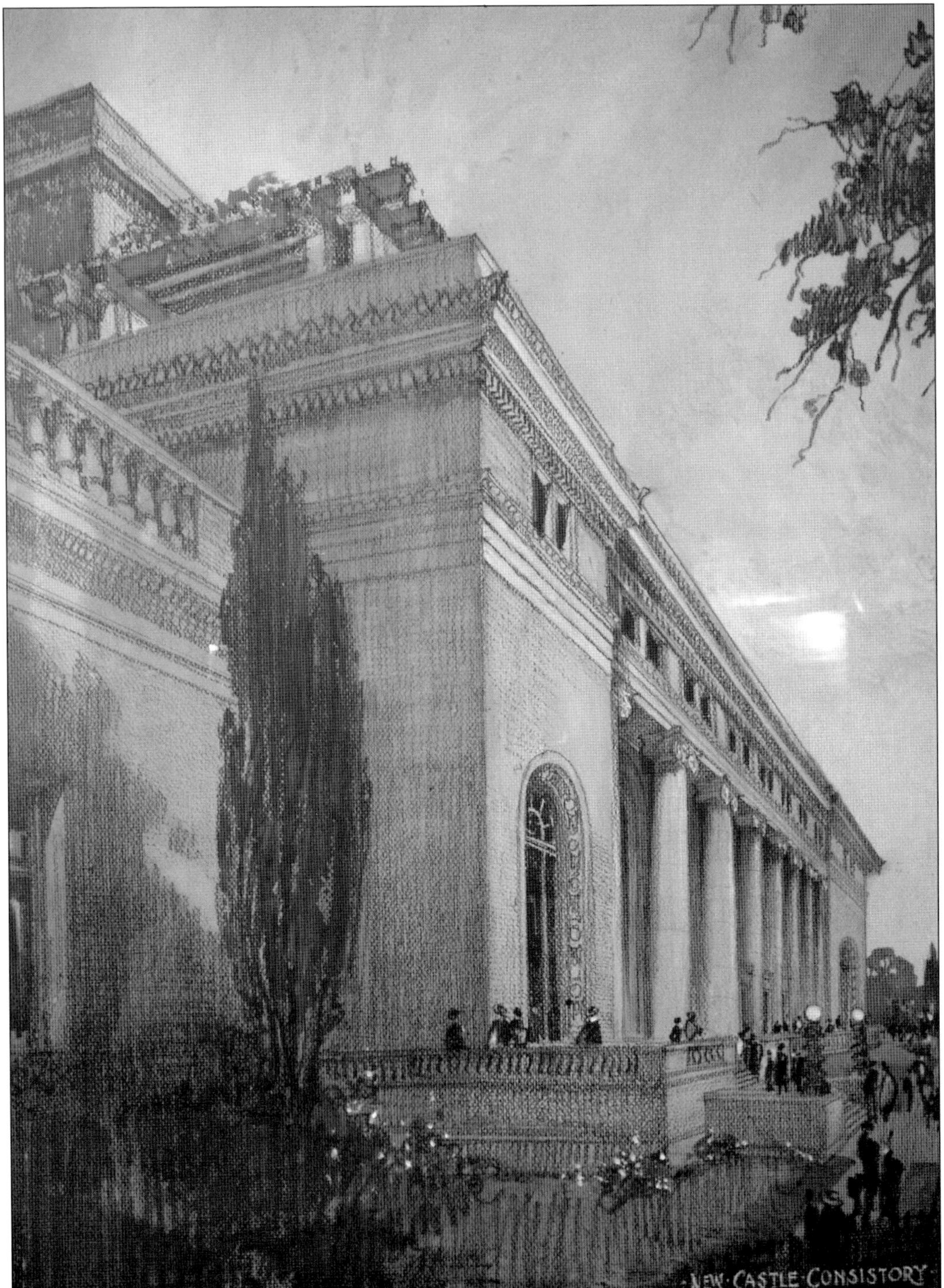

ARCHITECT'S RENDERING OF CATHEDRAL. Architect R.G. Schmidt drew up plans for New Castle's most impressive structure, bringing together Greek and Roman elements into a striking and timeless design. Budgetary restraints required the elimination of some of the details of this rendition.

One

Construction

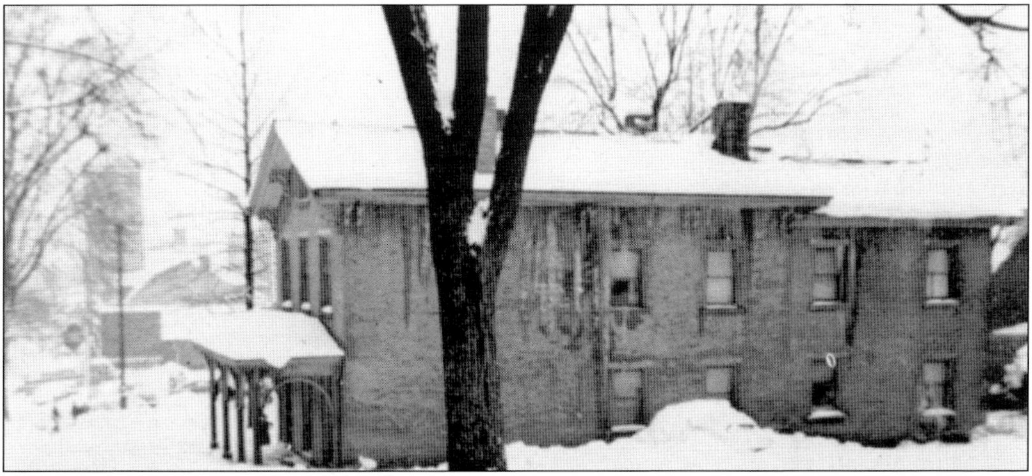

Crawford White Homestead. In July 1849 a group of Masons, primarily members from nearby Mercer Lodge No. 182, met secretly at the old Crawford White Homestead on the northwest corner of Jefferson and Grant Streets to petition the Grand Lodge of Pennsylvania for the constitution of Mahoning Lodge No. 243. This gathering apparently became the first documented meeting of Freemasonry in New Castle.

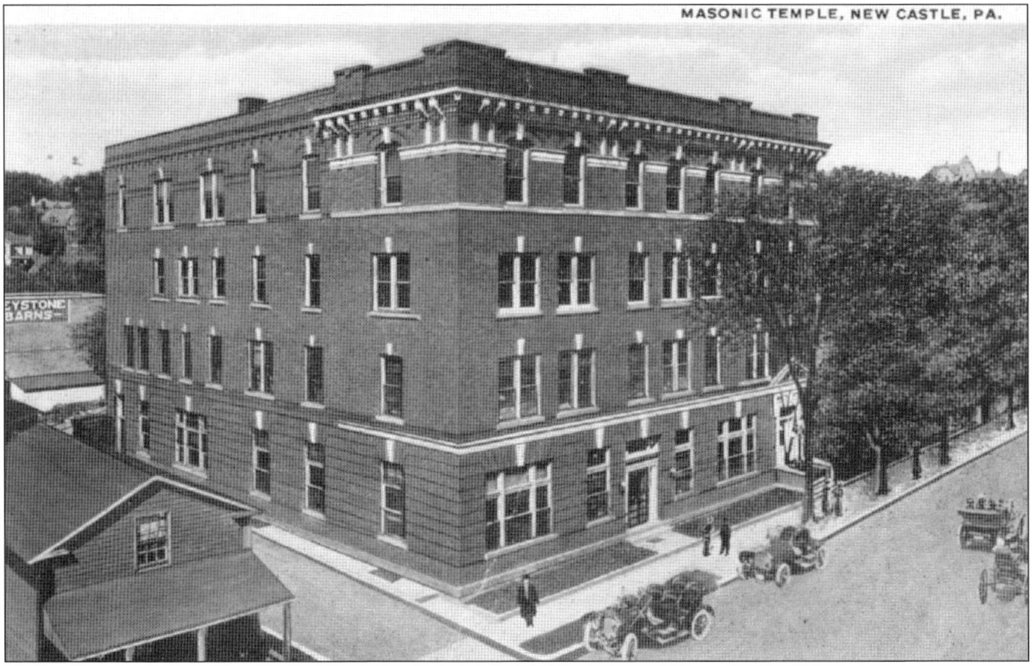

Masonic Temple Postcard. After moving to six different locations, New Castle's Masons constructed the Temple Building at 125 East North Street in 1905. Within two decades, the fraternal organizations had outgrown this facility, necessitating the building of the current Scottish Rite Cathedral, completed in 1926. Today, the Temple Building remains as an office and retail center.

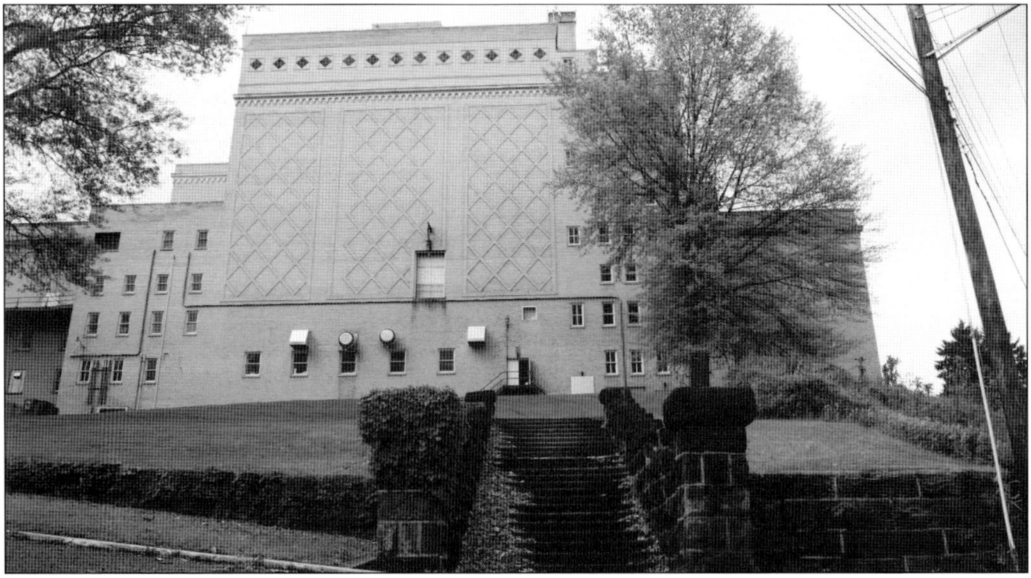

Stairway to Nowhere. Flush with hundreds of thousands of dollars, the Cathedral Association had the contractor install these steps supposedly leading to a massive front porch facing downtown. Unfortunately, when excavations revealed this area lacked the stability to support the weight of so large a structure, the architect altered the position of the building by rotating it 180 degrees, thus eliminating the need for this set of steps.

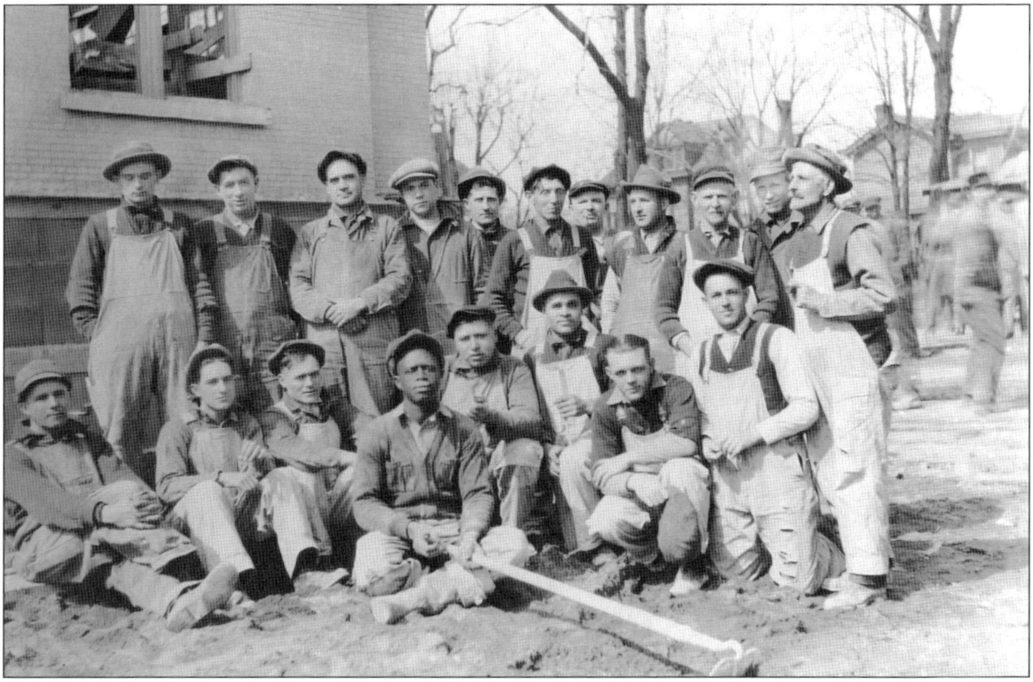

Construction Crew. These workers reflect a small sampling of the hundreds of steelworkers, bricklayers, electricians, plumbers, carpenters, plasterers, and painters required for the erection of the Scottish Rite Cathedral. Worker James R. Gilcrest is standing behind the mustached, tall man holding a pipe in the upper right-hand corner.

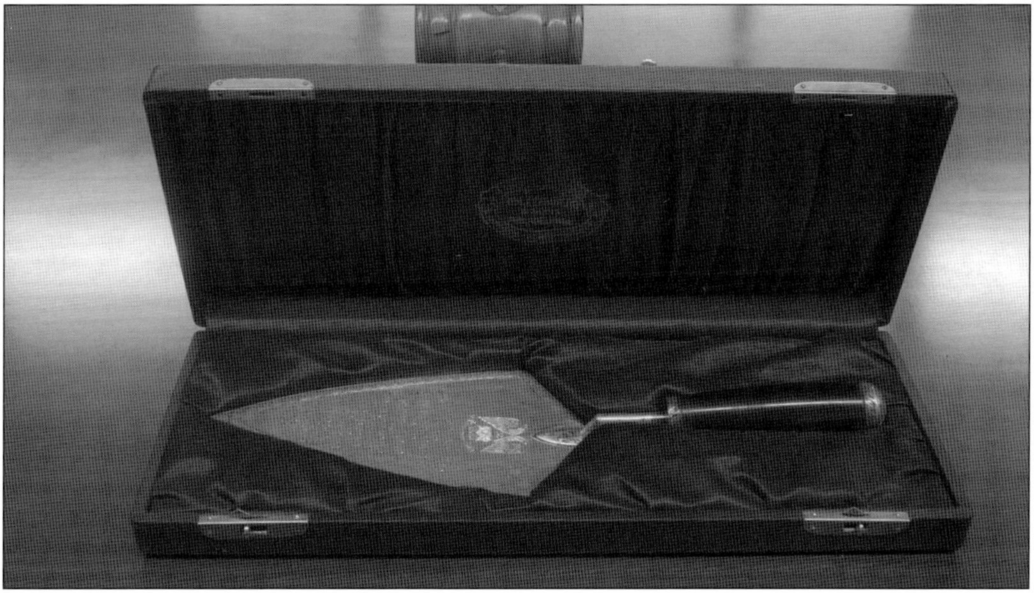

Workman's Trowel. James R. Gilcrest used his trowel to lay the first brick on November 13, 1924. The Right Worshipful Grand Master of the Grand Lodge of Pennsylvania, Samuel M. Goodyear, used this same trowel, which was later silver plated and engraved, to lay the cornerstone of the Cathedral at the formal ceremony held on June 10, 1925. Dignitaries received miniature trowels, donated by Brown & Hamilton Company as souvenirs.

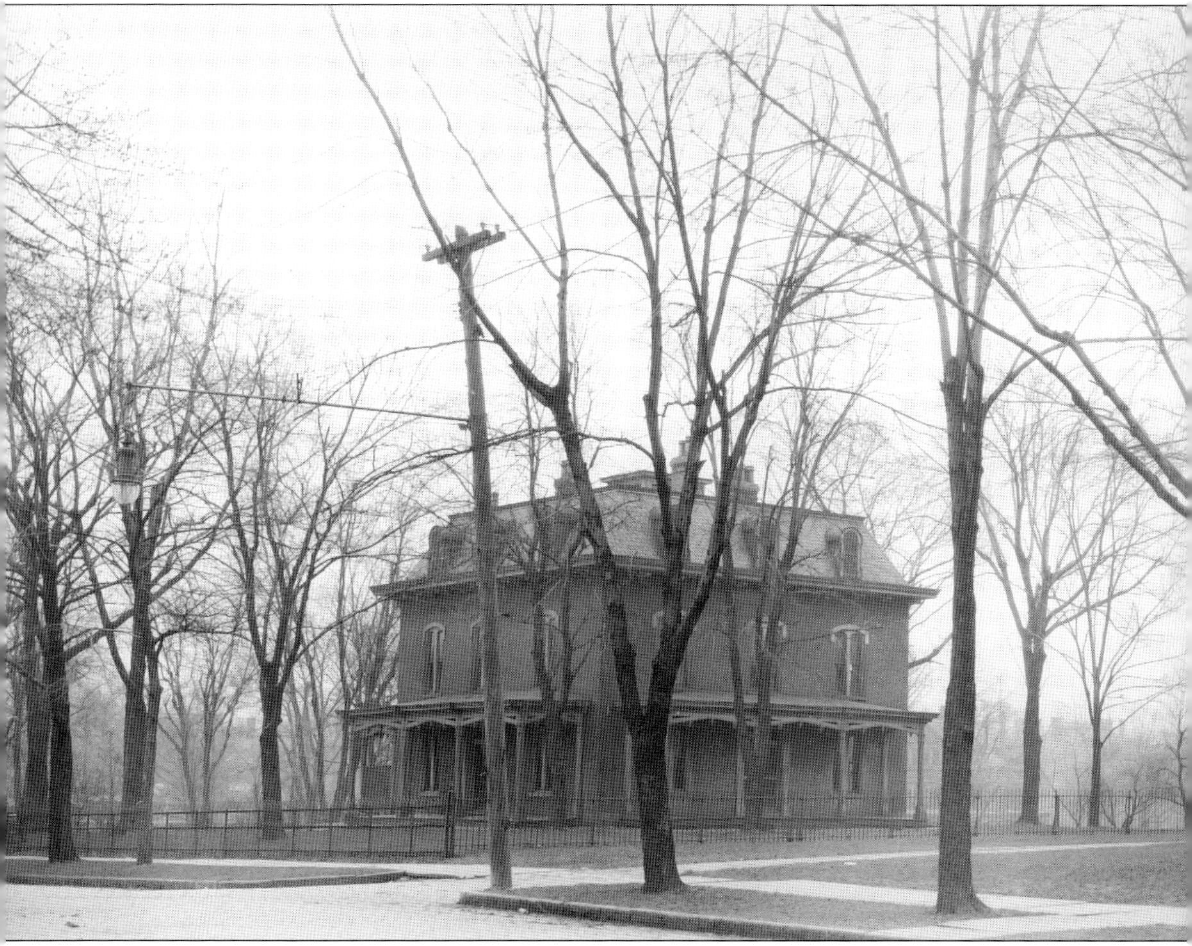

Knox Homes. Limited by the unstable ground at the lower level, the Cathedral Association relocated the construction site northwardly along Highland Avenue. The move required the demolition of the two stately mansions on Lincoln Avenue, formerly home to the Knox families. John Knox

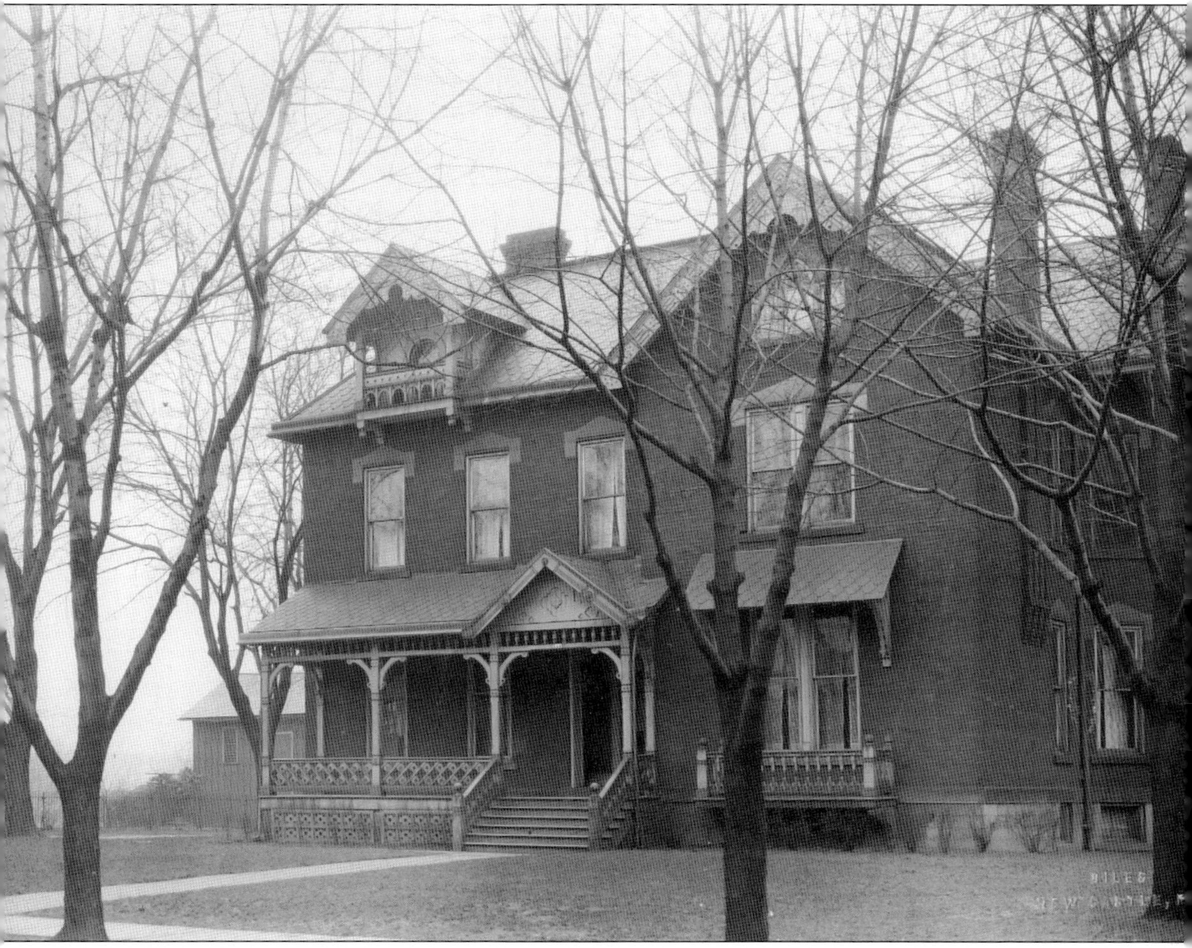

and his son ran New Castle's largest wholesale grocery company and retail store. Following the sale, the Knox families moved their residence farther north on Highland Avenue.

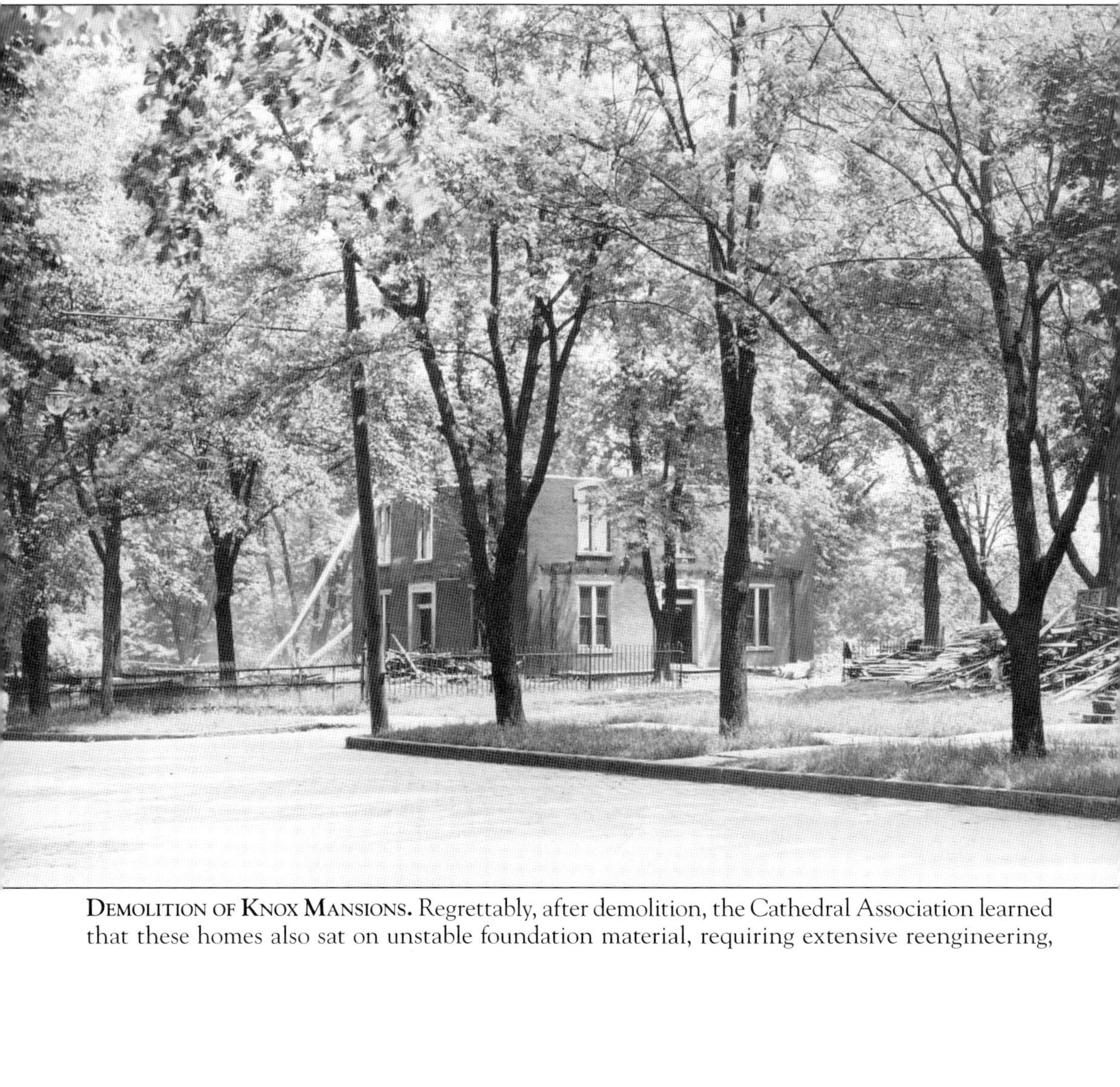

Demolition of Knox Mansions. Regrettably, after demolition, the Cathedral Association learned that these homes also sat on unstable foundation material, requiring extensive reengineering,

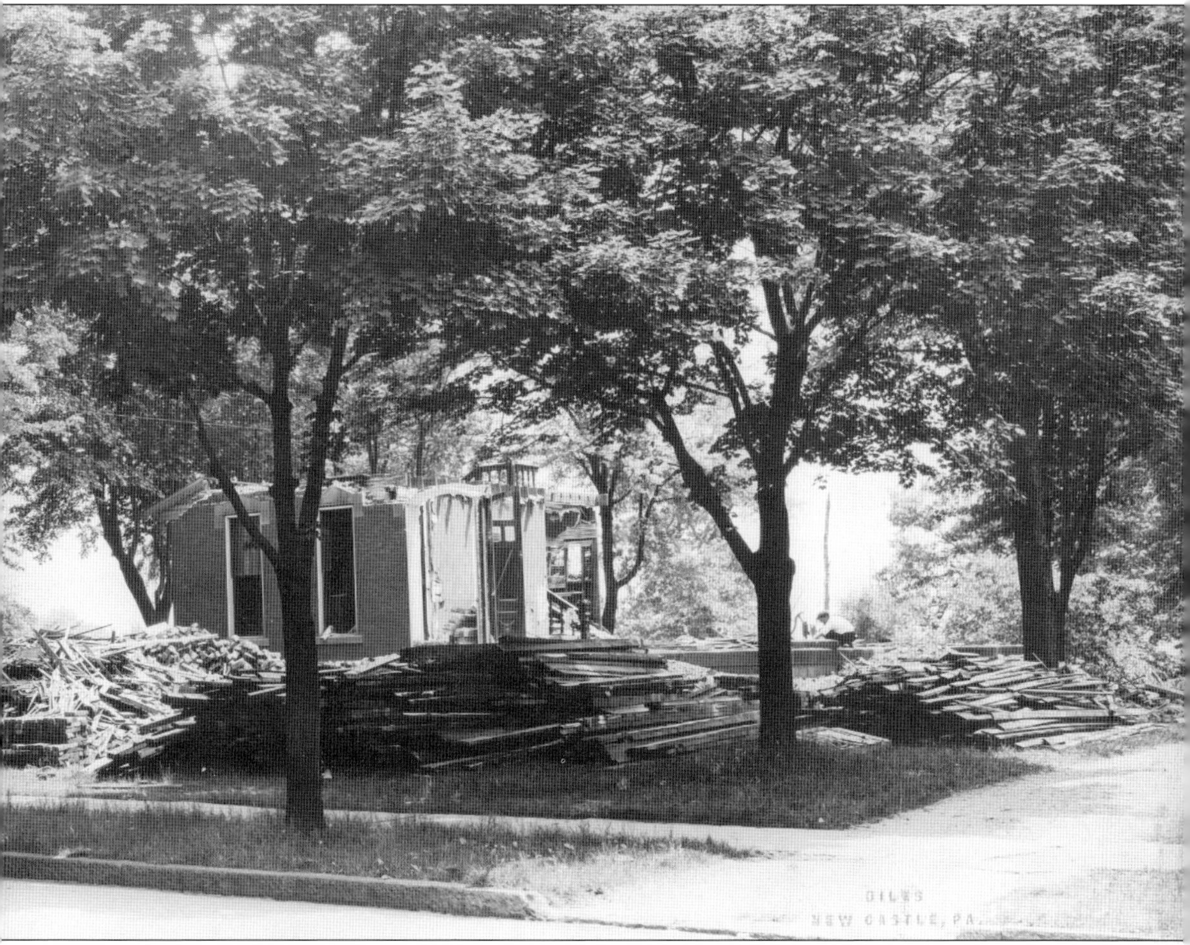

new architectural drawings, and the placement of nearly 1,000 support piles.

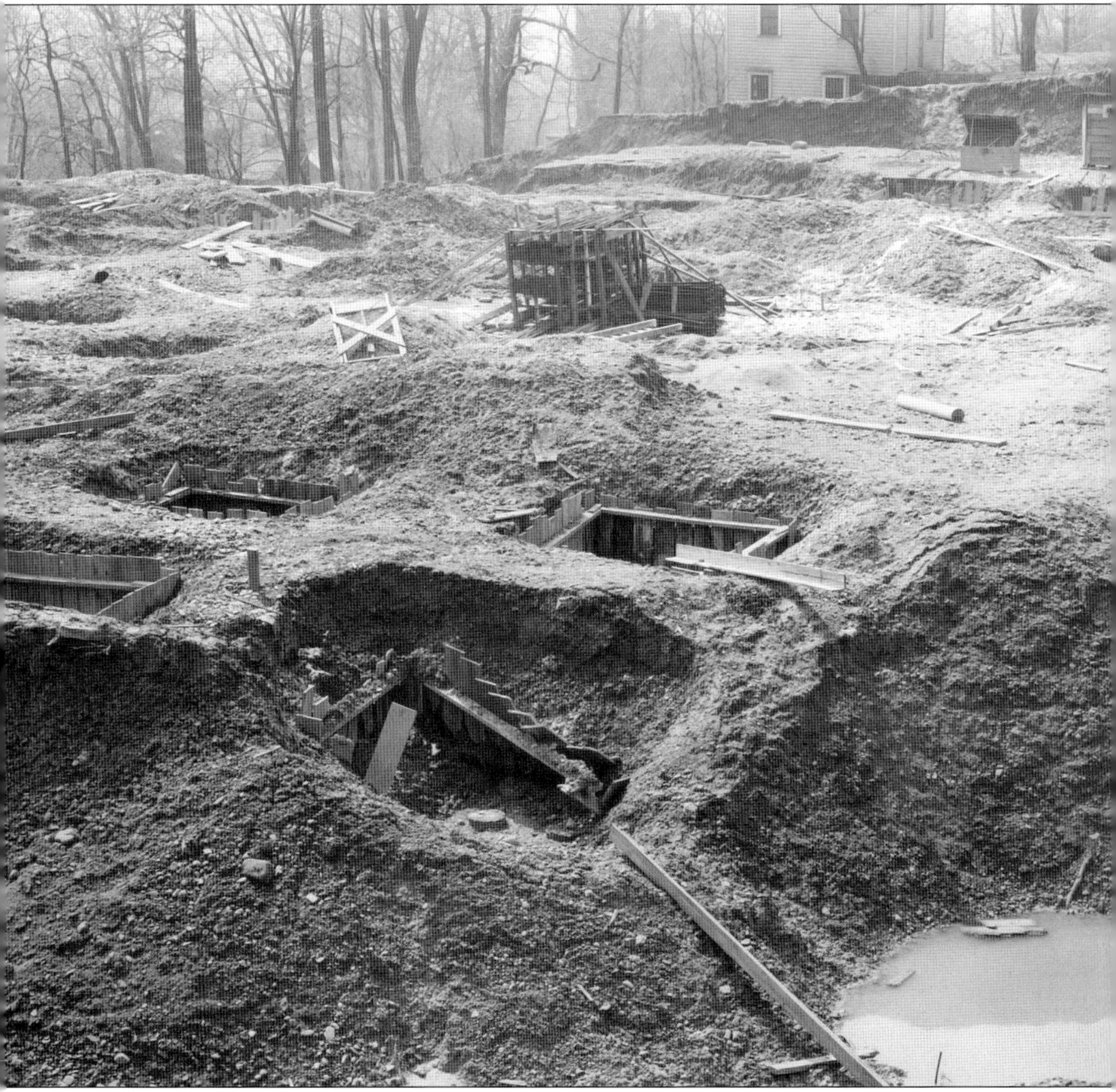

Installation of Pilings on Cleared Ground. The discovery of blue clay and sand forced the contractor to install 916 support piles at depths of 20 to 26 feet, costing $32,000, almost $450,000

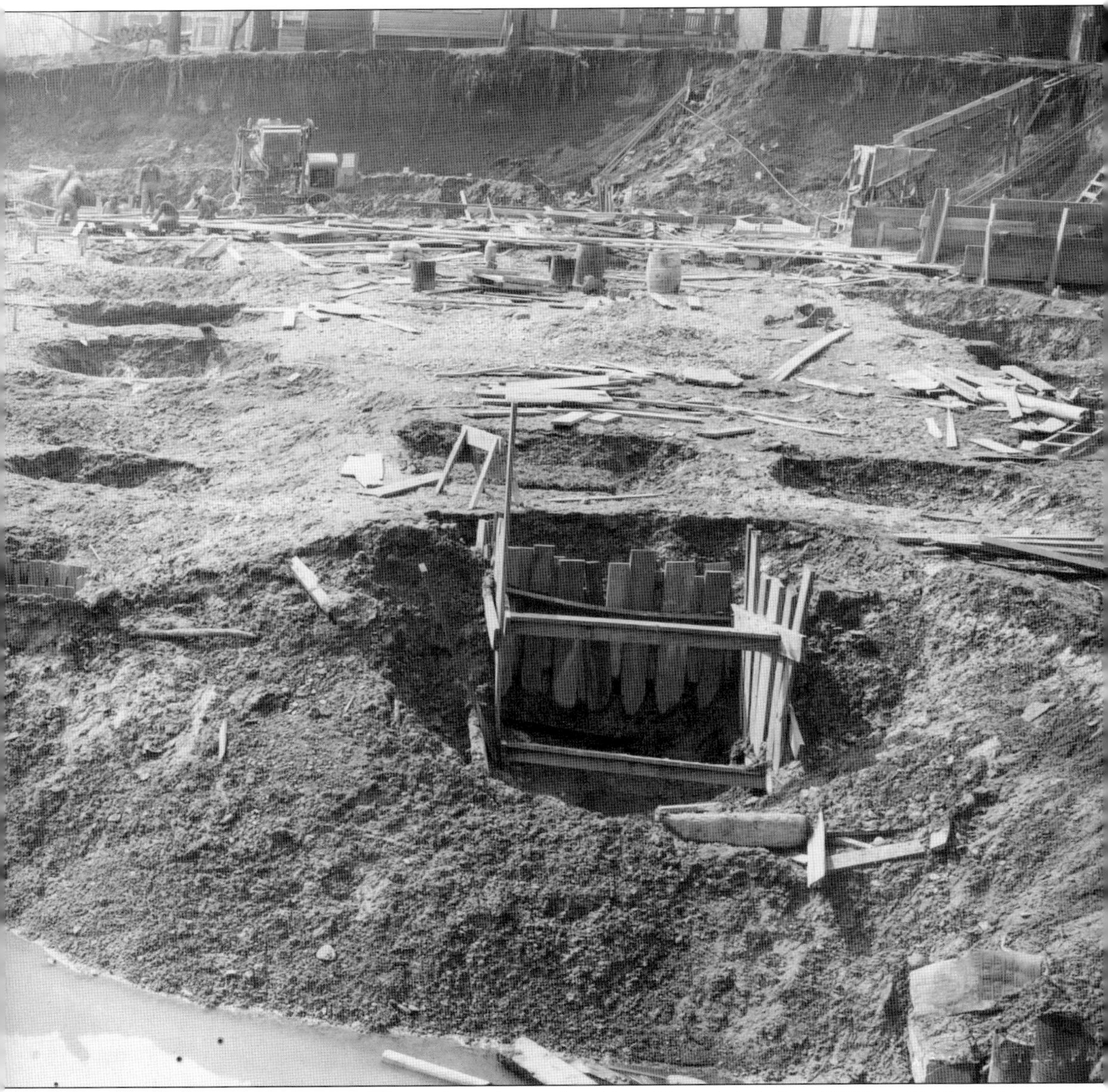

in current dollars. Former foundation board president Ken Shiderly's father, William E. Shiderly, helped dig out and haul away the unused fill, clearing space for construction.

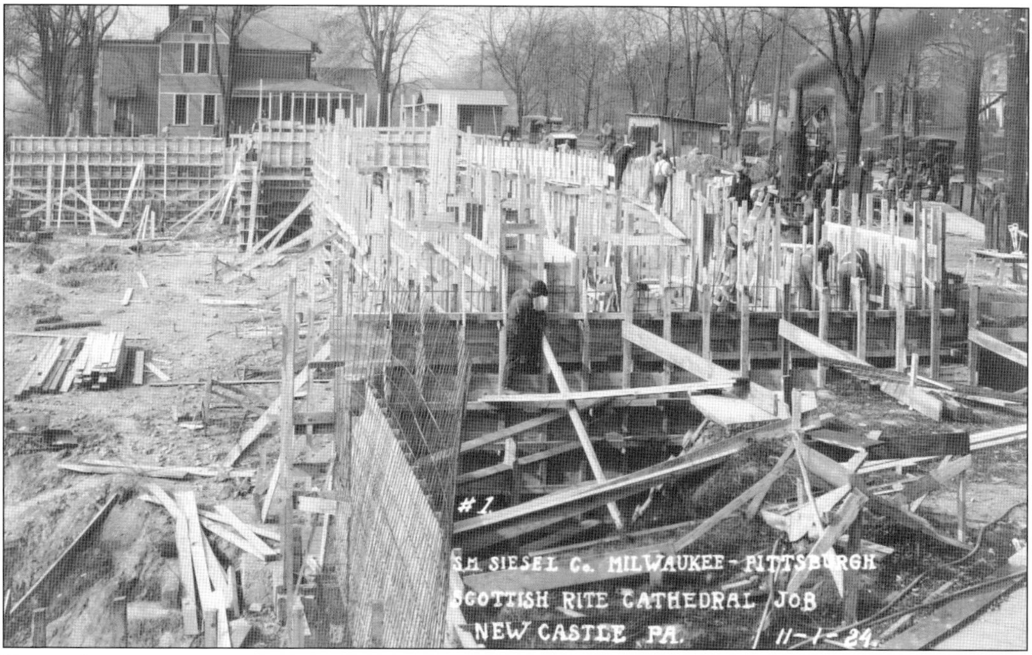

EARLY CONSTRUCTION, NOVEMBER 1, 1924. Wooden forms begin shaping the Cathedral. Dozens of carpenters nail support beams, giving life to the foundation of the building. The large home visible to the right side belonged to the Miller family and later would become part of the west parking lot.

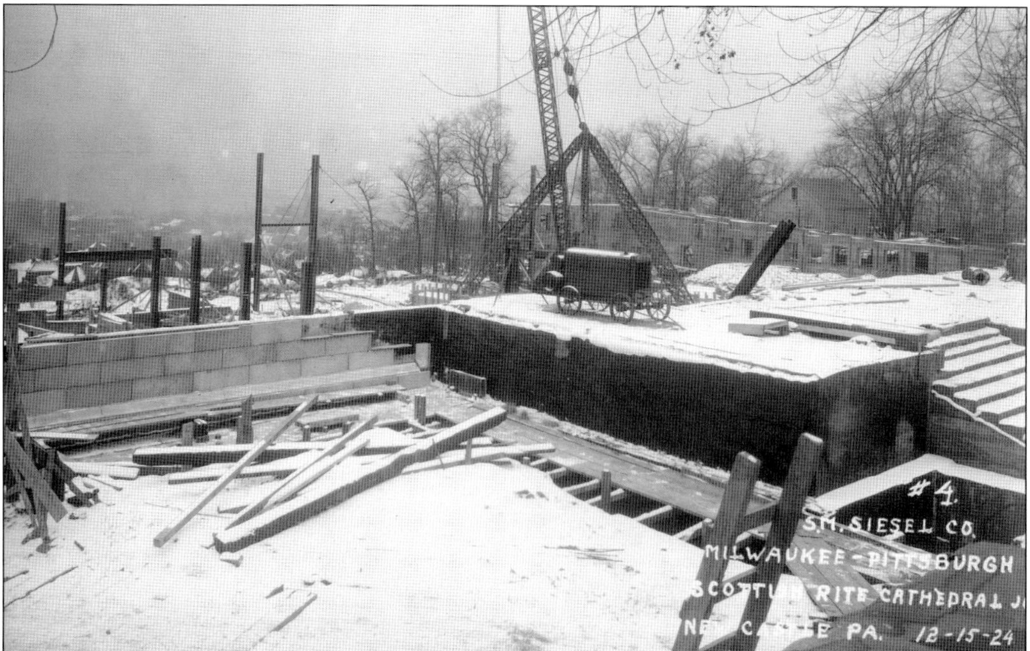

CONSTRUCTION, DECEMBER 15, 1924. An Ingersoll-Rand steam generator supplied power to the site for the operation of a giant crane. The equipment hoisted beams in frigid winter temperatures as low as four degrees below zero.

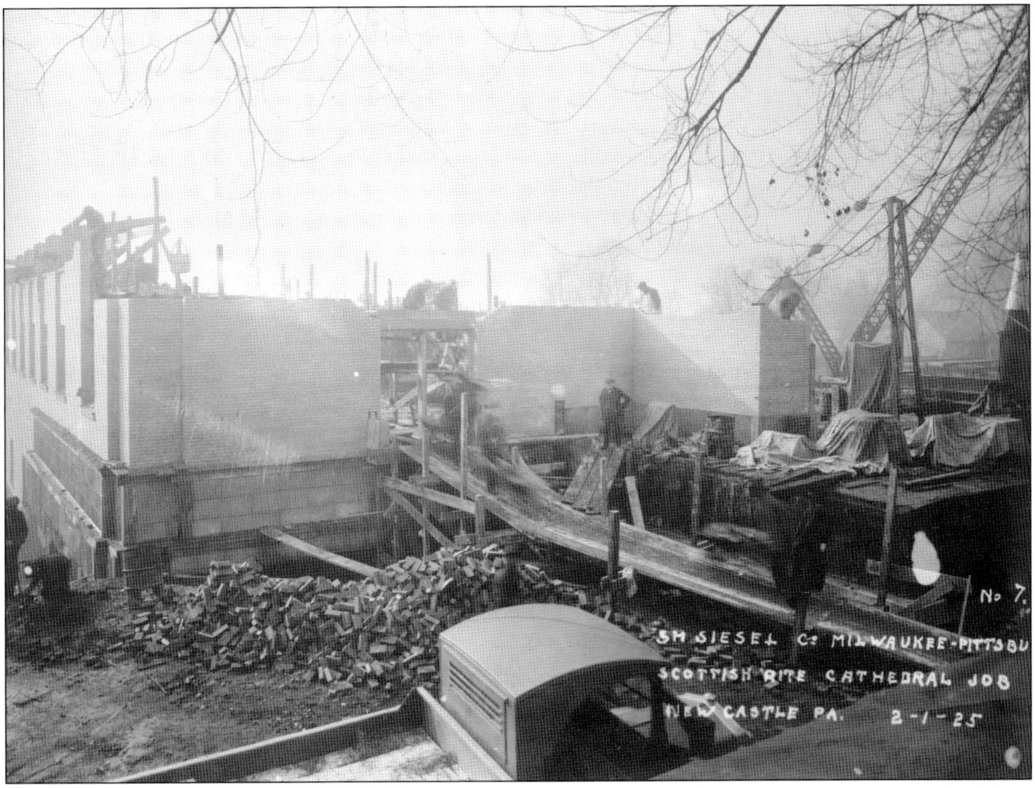

CONSTRUCTION, FEBRUARY 1, 1925. A truck delivers brick to the stonemasons as the building takes shape. Note the walkway where tenders wheeled all bricks by hand. In the background, the giant crane lifts larger materials such as steel beams and girders.

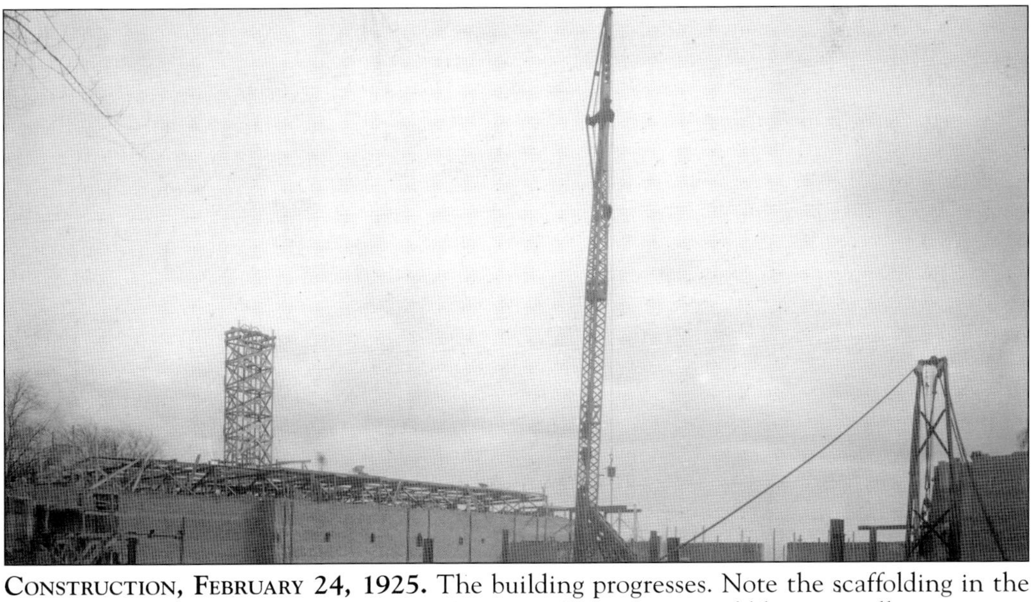

CONSTRUCTION, FEBRUARY 24, 1925. The building progresses. Note the scaffolding in the background. The brickwork rises skyward. Eventually, masons would lay two million common bricks and 600,000 decorative exterior-facing bricks to complete the Cathedral.

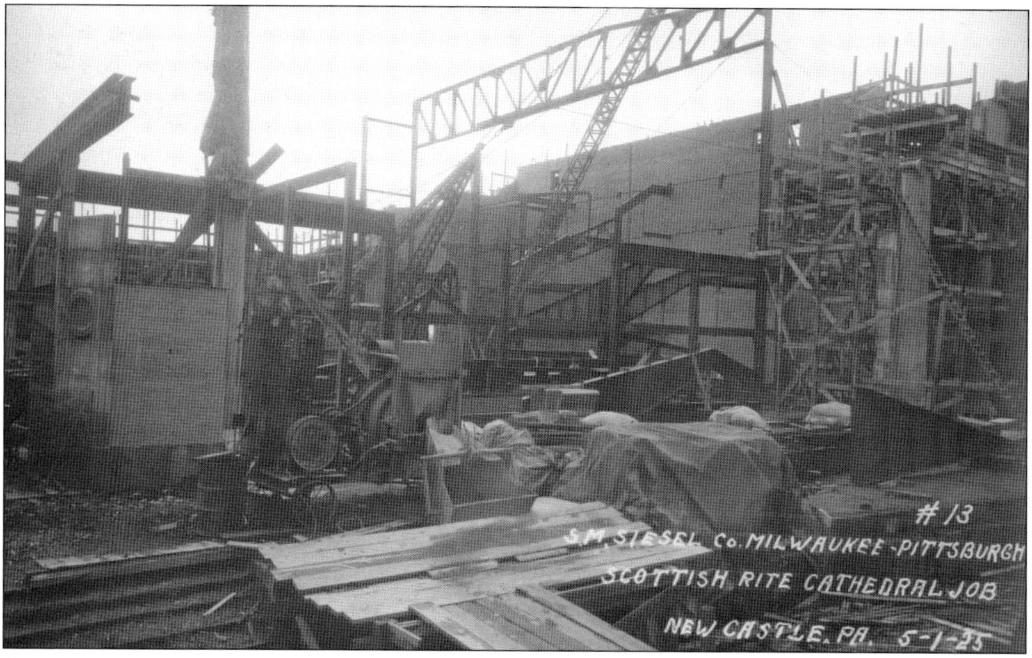

Construction Continues, May 1, 1925. Beams rise, walls grow, and scaffolding supports 32,000-pound Indiana limestone columns adorning the front of the building. Amazingly, the construction moves ahead with incredible speed.

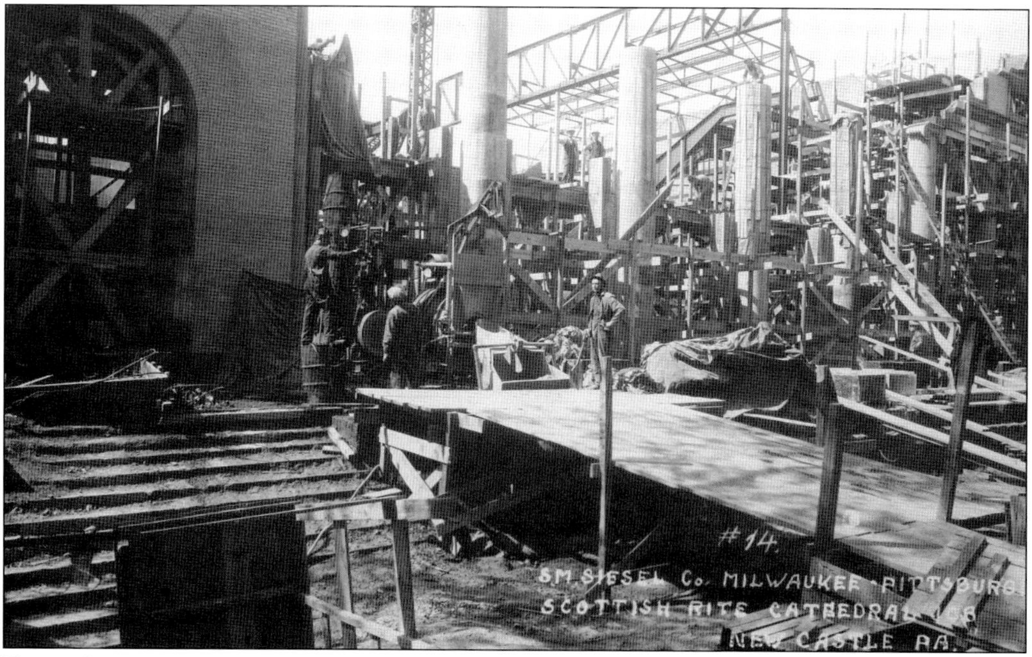

Front Construction, May 15, 1925. Workers position eight Greek-inspired, 32-foot-tall (when pieced together), 32,000-pound columns from the Indiana Limestone Company to span across the front of the building. Teamsters haul each section by horse and wagon up the steep Jefferson Street hill from the train station to the work site. Roof supports for the lobby and auditorium take shape in the background.

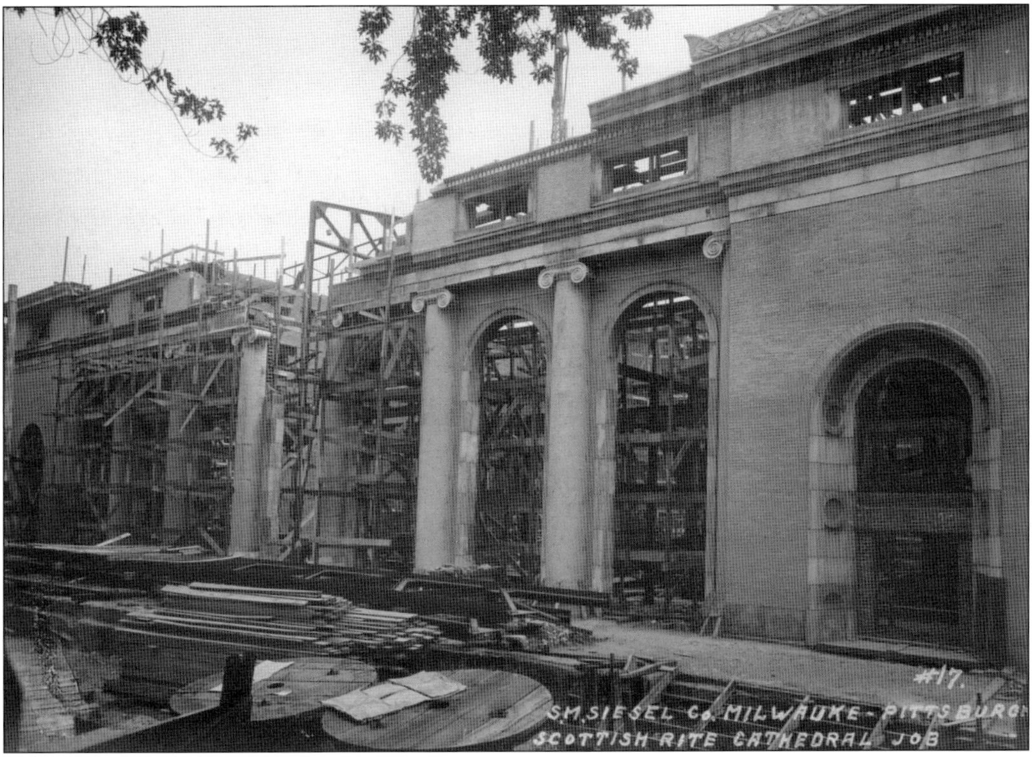

Exterior Nears Completion, July 1, 1925. After less than nine months of construction, using primitive building techniques by today's standards, lacking fork-trucks, automation, drones, and electronically controlled tools, the building exterior takes final shape. Skilled artisans assisted by day laborers performed most work by hand.

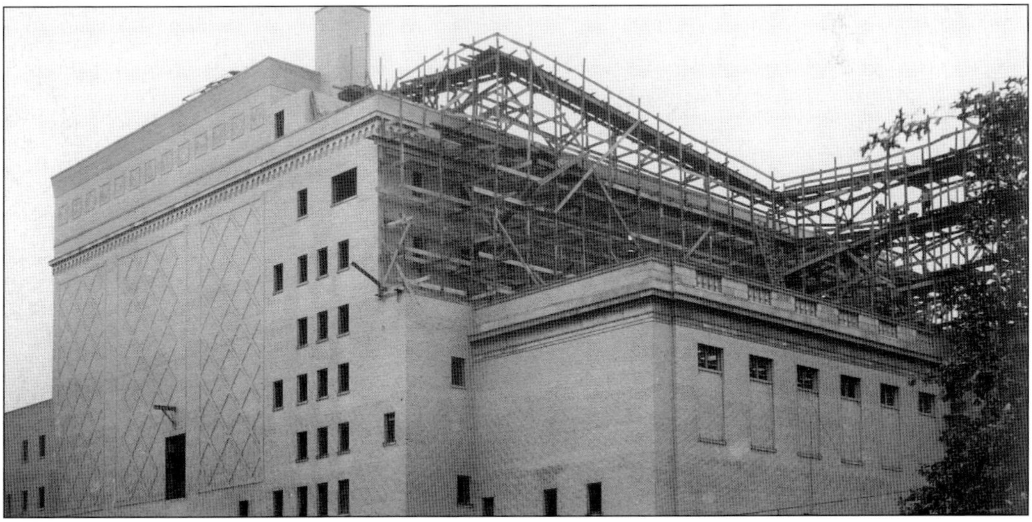

Finishing Touches, August 1925. A patchwork of scaffolding enables the bricklayers to apply the last of the 600,000 finishing bricks used throughout the face of the building. This rear view of the Cathedral highlights the geometric and lattice-work designs, plus the Greek-style, via-and-mutual border, a series of grooves and extensions to add dimension, employed across the roofline of the ninth-floor edifice.

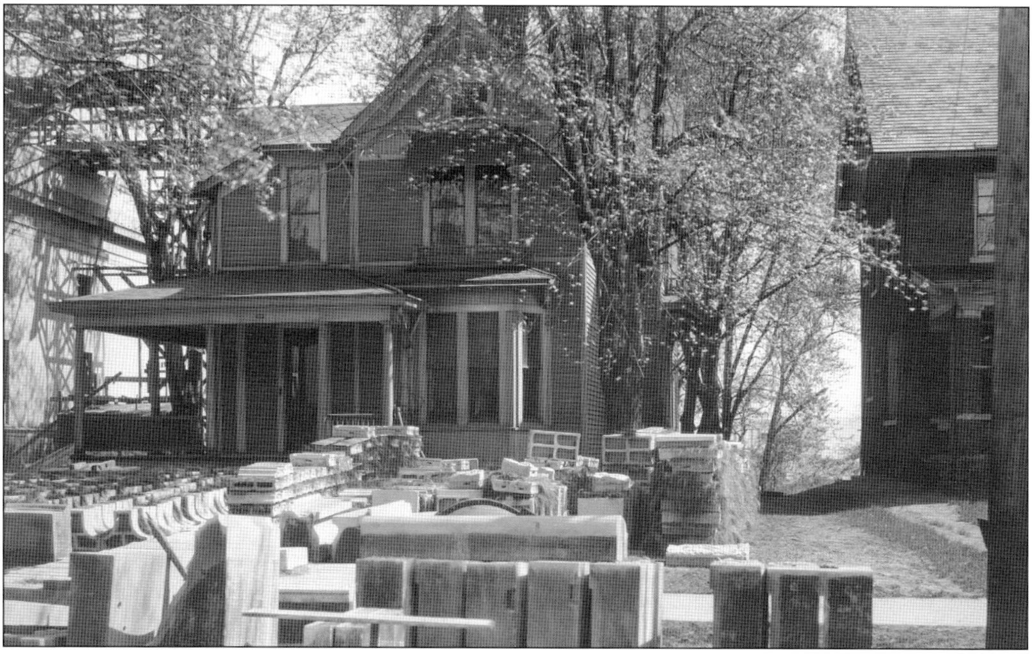

Homes Scheduled for Demolition. Indiana limestone capstones line the front yard of the J.R. Miller home neighboring the Cathedral. The corner of the John S. Wallace house peeks out from the right side of the photograph. A demolition crew would eventually raze both homes to make way for the west parking lot.

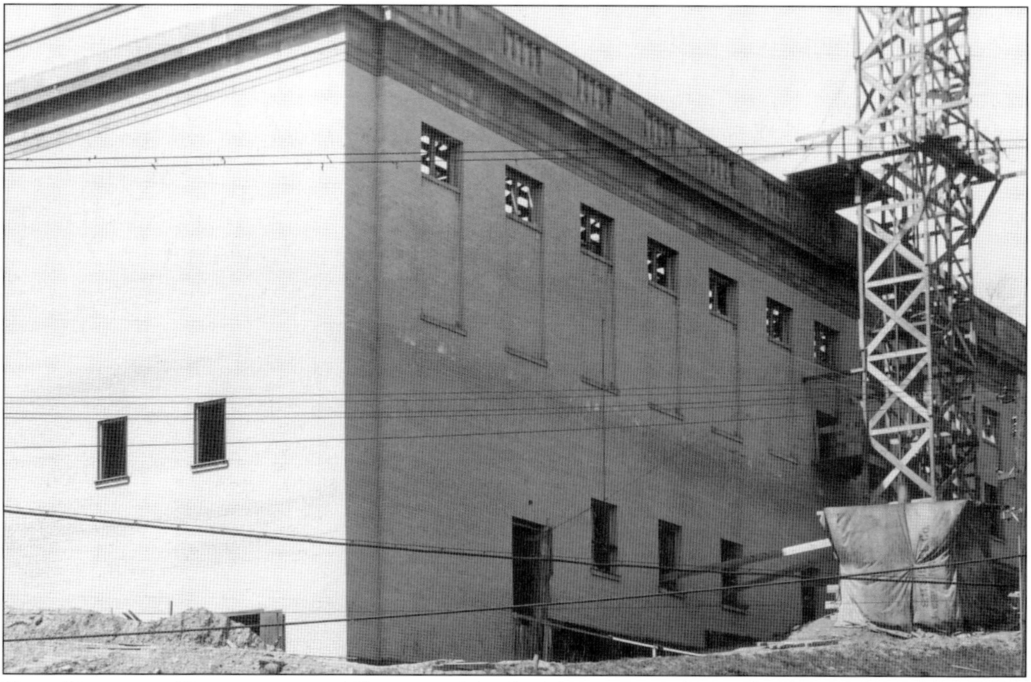

Rear Building Construction. The rear of the building moves skyward. The section on the right side of the photograph shows the East Lodge Room. The West Lodge Room stands identical in shape and size to the East Lodge Room and remains visible on the left side of the photograph.

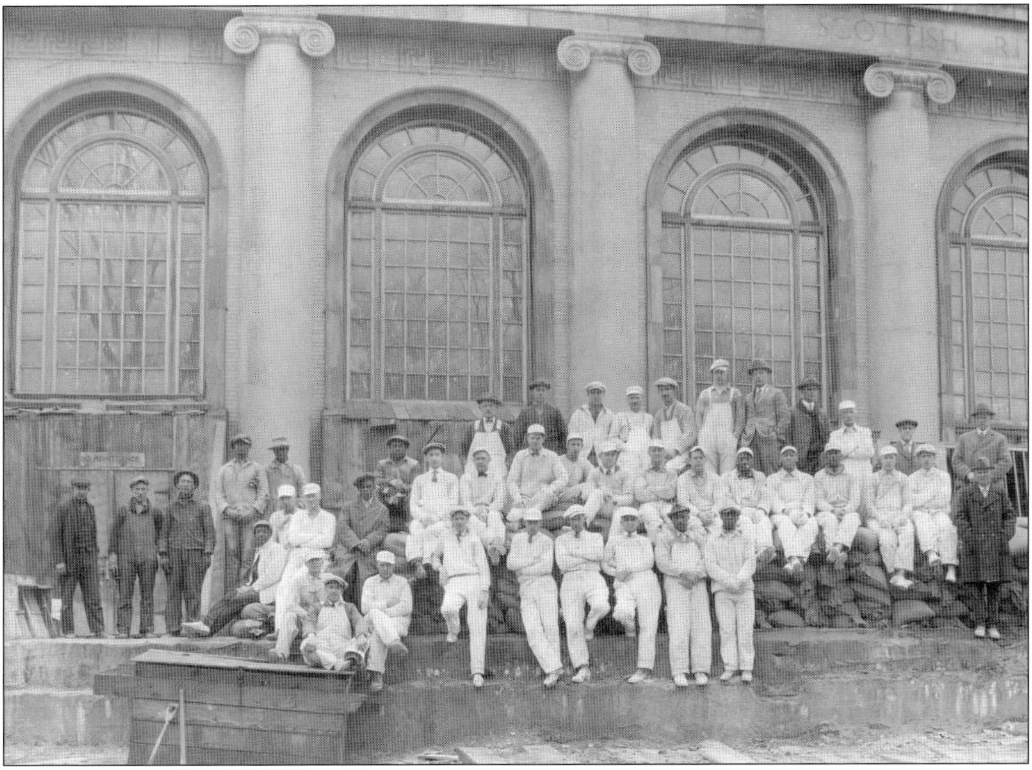

Workmen on Extra Break. Dozens of workers pose for a photograph while standing on the soon-to-be-completed front porch of the Cathedral. These craftsmen displayed a high level of skill and speed while creating New Castle's signature structure.

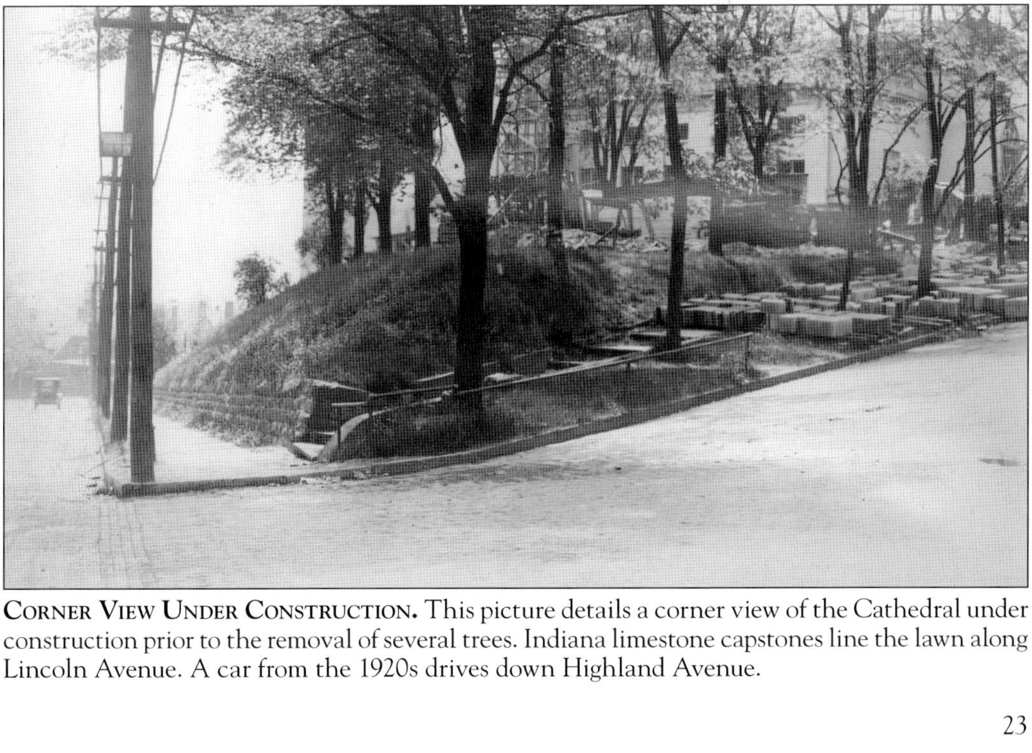

Corner View Under Construction. This picture details a corner view of the Cathedral under construction prior to the removal of several trees. Indiana limestone capstones line the lawn along Lincoln Avenue. A car from the 1920s drives down Highland Avenue.

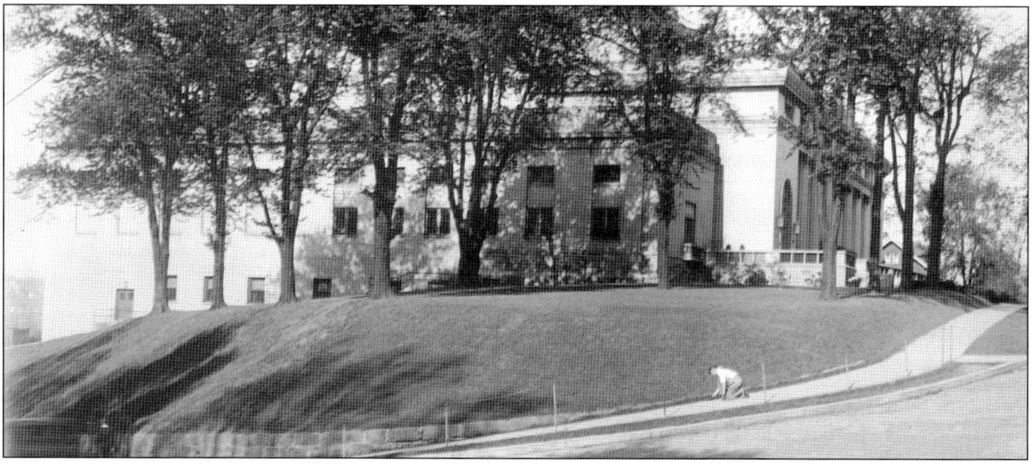

Corner View of Cathedral. An operative mason adds the finishing touches to the sidewalk at the corner of Lincoln and Highland Avenues. Most of the landscaping has been completed. The stone retention wall along the hillside remains in place today.

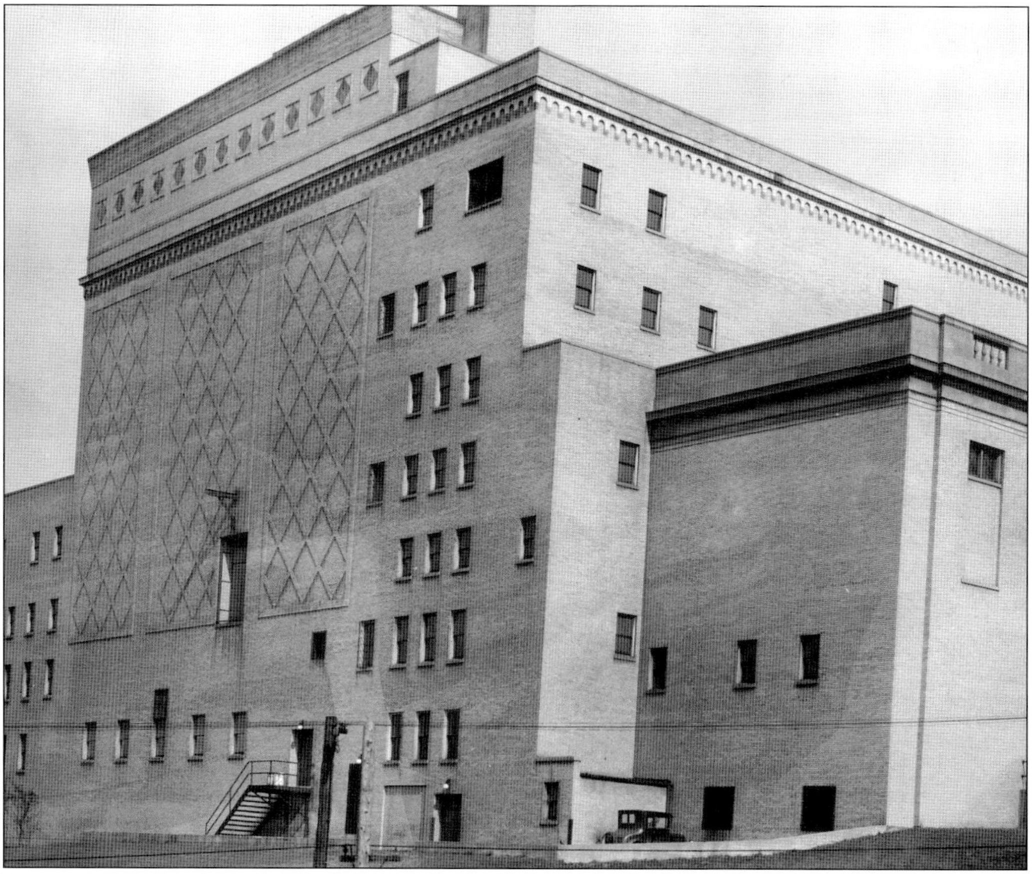

Rear and East Side. This photograph details some of the architectural highlights found throughout the structure of the building. Greek and Roman adornments add richness to the exterior of the Cathedral, testifying to the excellence of the architect who designed it and the operative masons who constructed it.

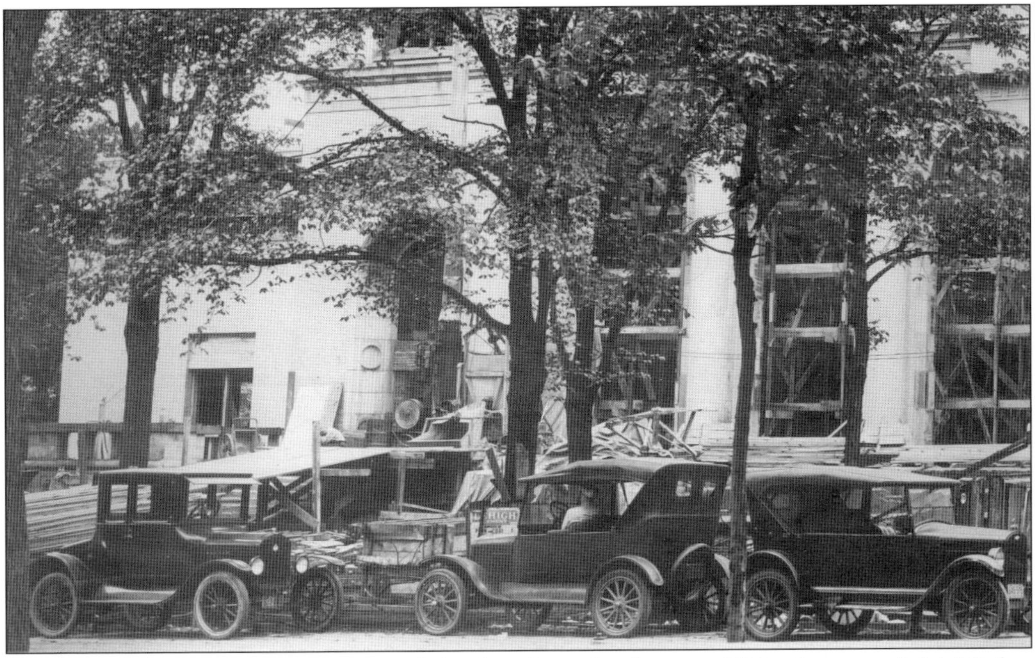

Craftsmen's Period Cars. Automobiles line Lincoln Avenue. This photograph highlights some of the classical detail on the building exterior. Classically influenced vias and mutuals, a traditional Meander design, named for the ancient Meander River in Greece, and scrollwork above the columns add impact and originality to the Cathedral.

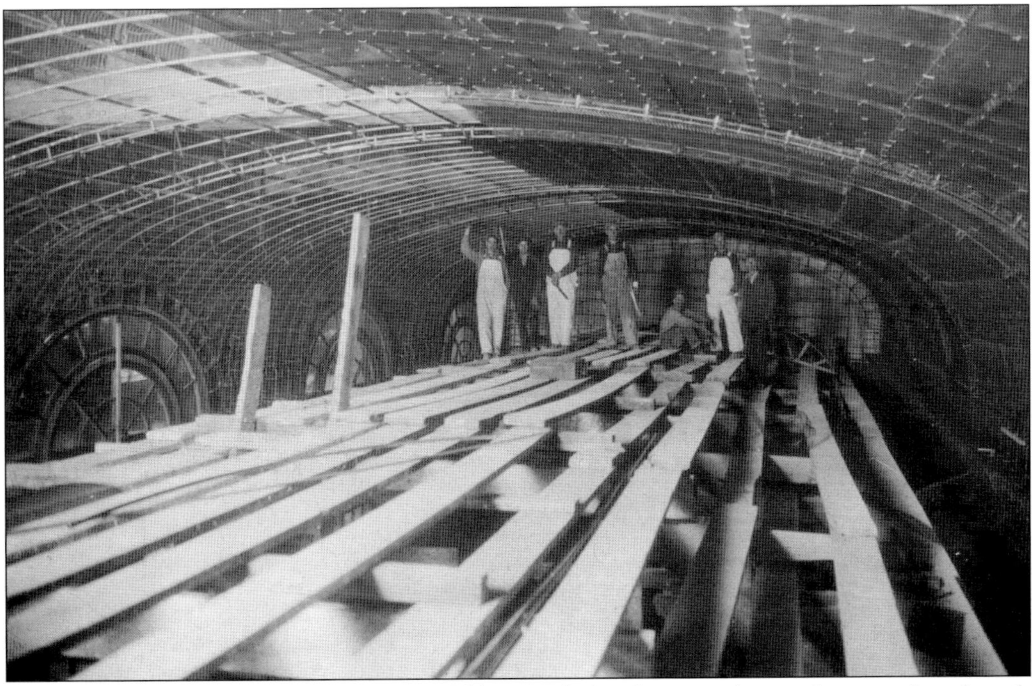

Lobby Construction. Workmen on scaffolding apply finishing touches to the curved steel-mesh ceiling of the lobby prior to plastering. Seven 9-by-15-foot arched windows, each with 36 individual panes, add a perfect balance to the room design and the front of the building.

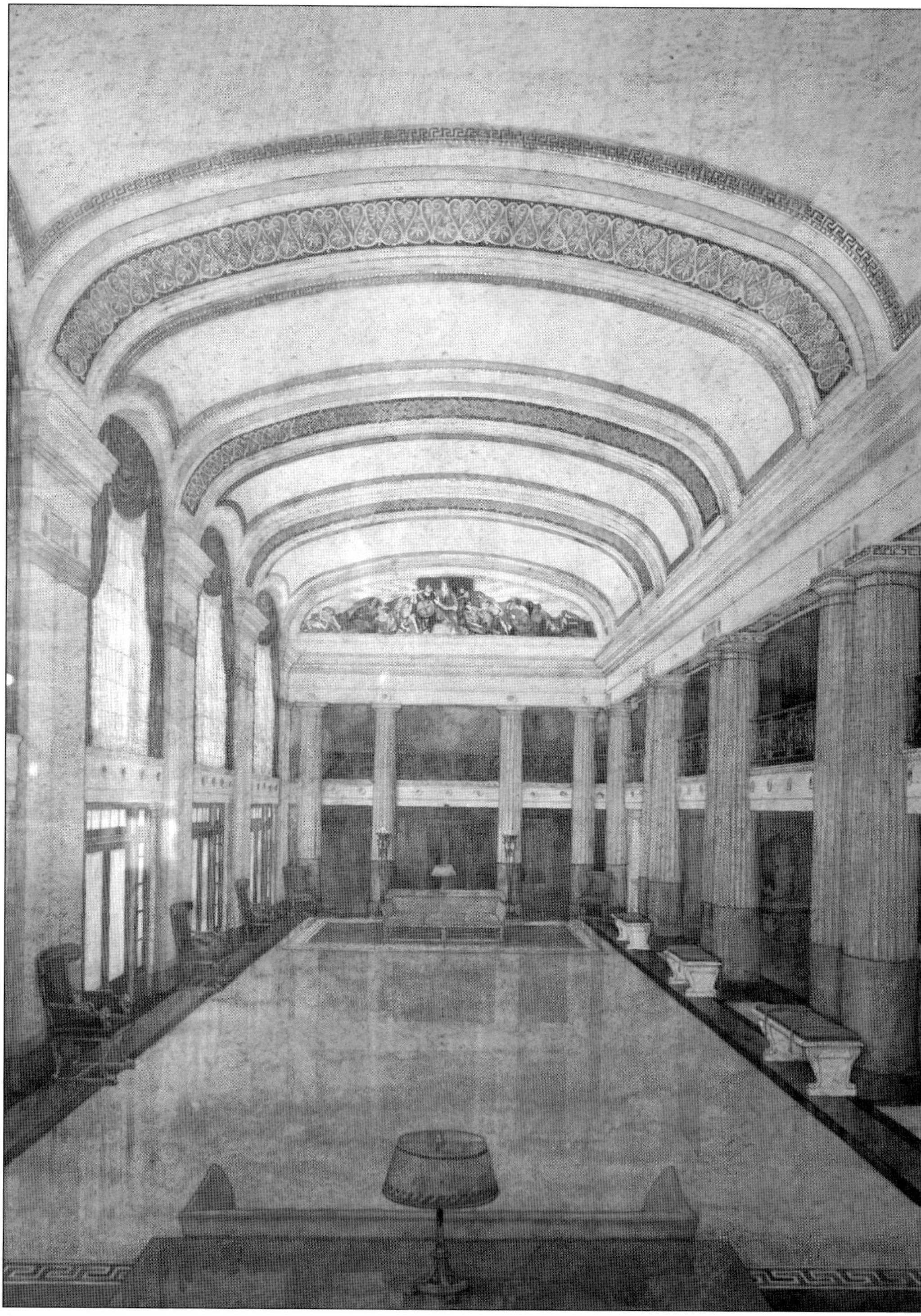

ARTIST RENDITION OF LOBBY. Although budgetary considerations required cutbacks in some of the details, this rendering portrays a very close approximation of how the Cathedral lobby actually looks today, minus the mural, curtains, and benches.

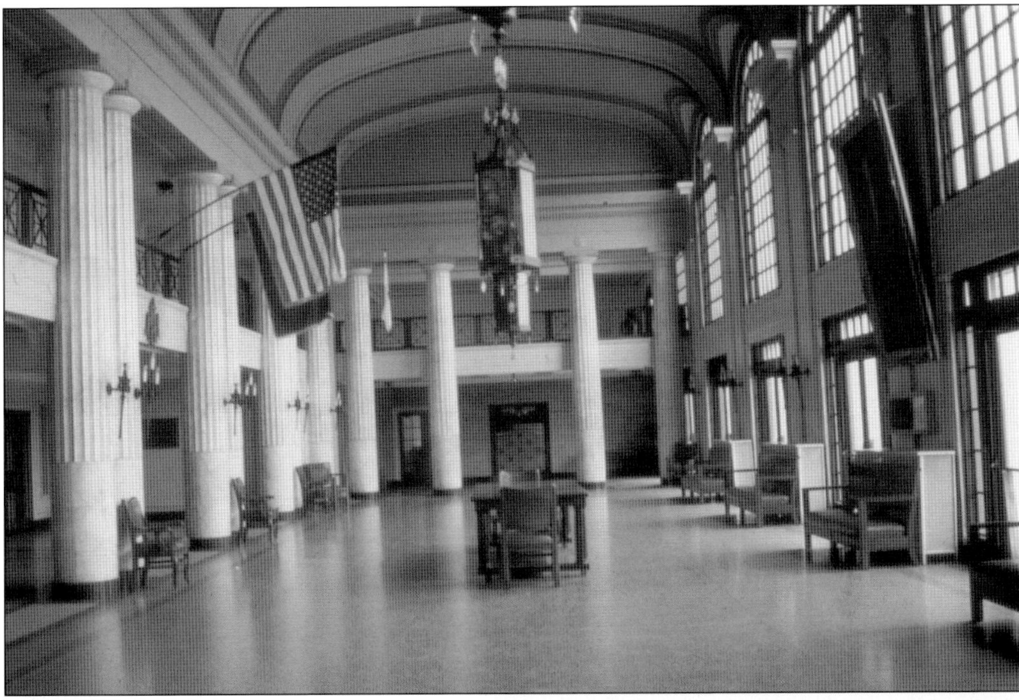

CURRENT LOBBY VIEW. Often compared to the temple of Diana in Ephesus, the 70-by-40-foot lobby may represent the signature room in the Cathedral. The three massive chandeliers, the rows of 18 cream-colored, mottled Greek Revival columns, 20 wrought-iron sconces, and the decorative mezzanine railing add refinement to an area brightly lit with the natural light from seven immense windows. A Meander pattern tops the pillars and highlights the ceiling.

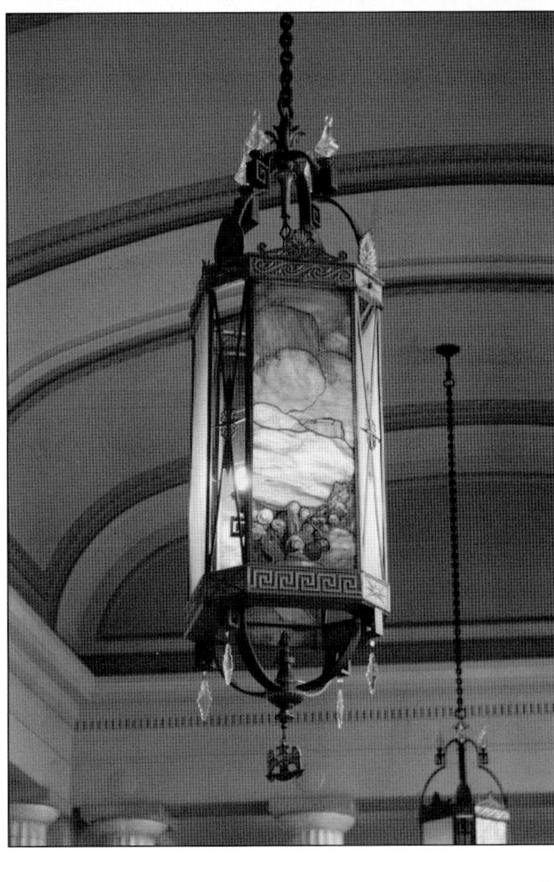

LOBBY CHANDELIERS. A heavy black-link, 10-foot chain suspends each of the three chandeliers from the 45-foot tall ceiling. Two sides of each chandelier contain frosted glass panels, and another two contain unique stained-glass scenes from Consistory degree work. Four crystal finials top each chandelier while four crystal drops and a bronze double-headed eagle hang from the bottom. Decorative metalwork borders the glass panes, which measure more than three feet in length.

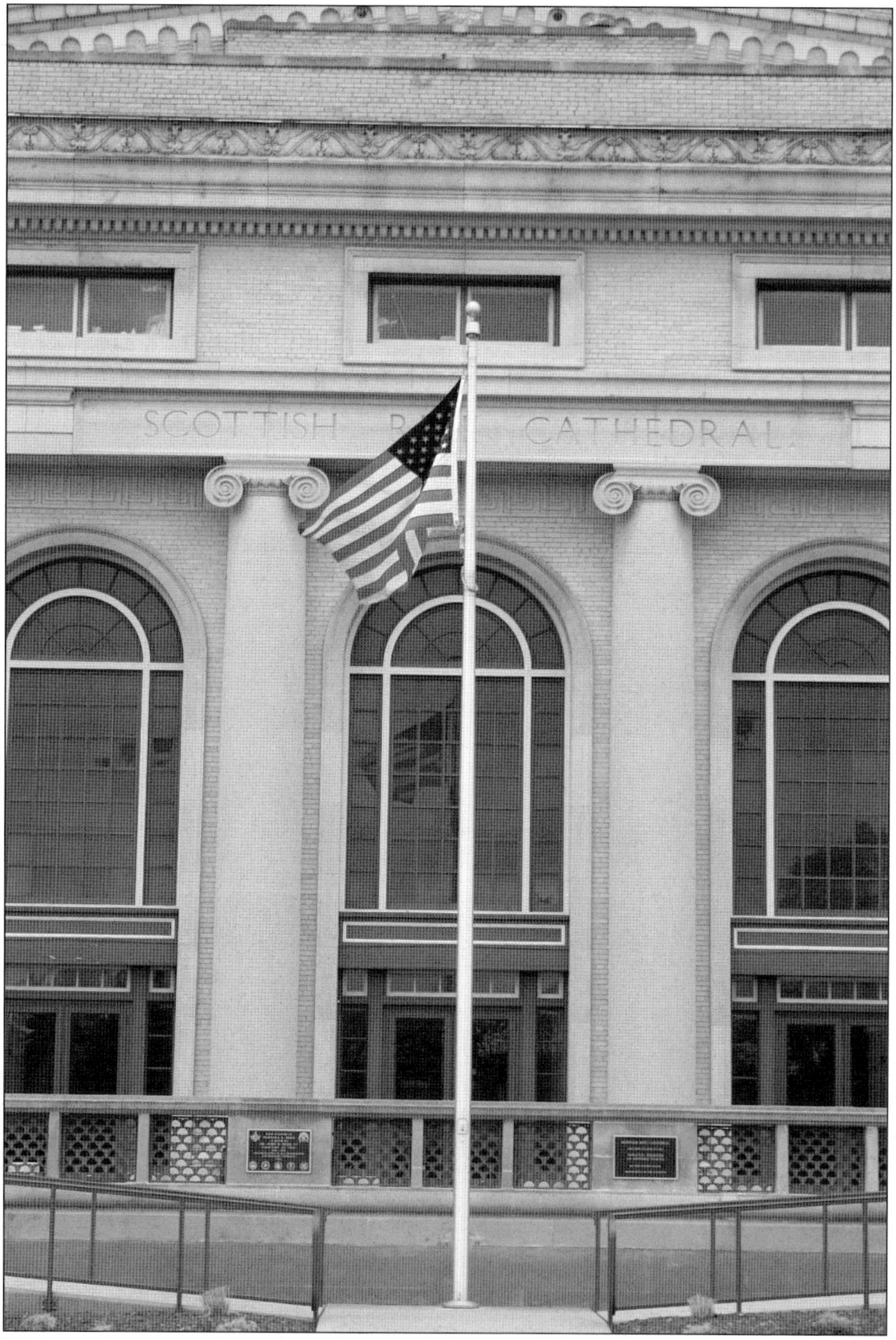

West View of Cathedral Front. The roomy front porch frequently houses weddings ceremonies, receptions, reunions and other events. The seven magnificent front windows and doors pour light into the lobby. The Cathedral measures 245 in width by 181 feet in depth and 110 feet in height, and the front terrace measures 170 feet in width by 24 feet in depth. The building contained 880 electrical outlets and more than 2,000 lamps at the time of construction. The Cathedral claims 210 windows and 436 doors. The contractor used 1,400 tons of steel and 36,000 sacks of cement. The edifice required 2.6 million bricks.

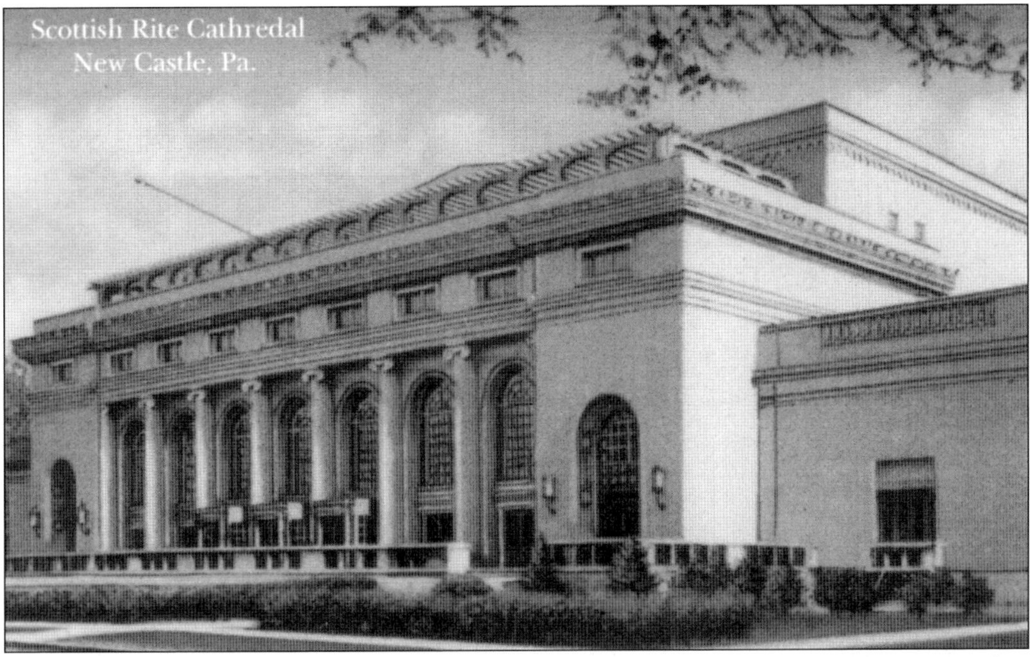

POSTCARD OF FINISHED CATHEDRAL. This picture dramatizes the impressive 168,271-square-foot structure currently valued by Industrial Appraisal Company at a replacement value of $67,285,410. The building sitting at the corner Highland and Lincoln Avenues stands as a crowning jewel for all New Castle to see and admire.

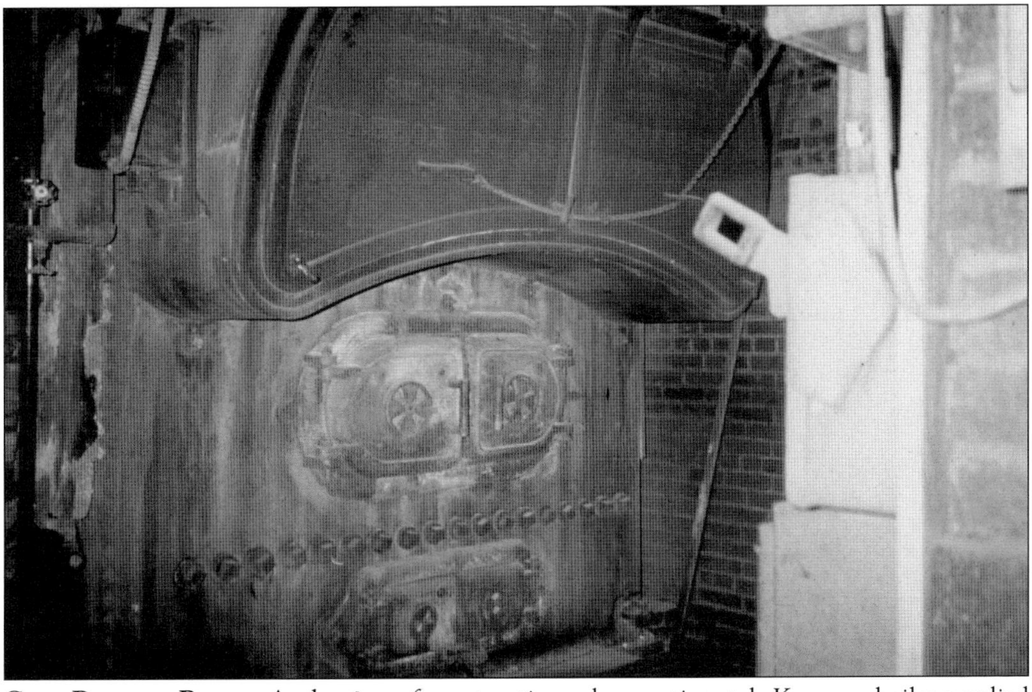

COAL-BURNING BOILER. At the time of construction, a locomotive-style Kewanee boiler supplied heat for the Cathedral. Over the years, coal has given way to natural gas, and a modern H.B. Smith Company of Westfield, Massachusetts, cast-iron system has replaced it.

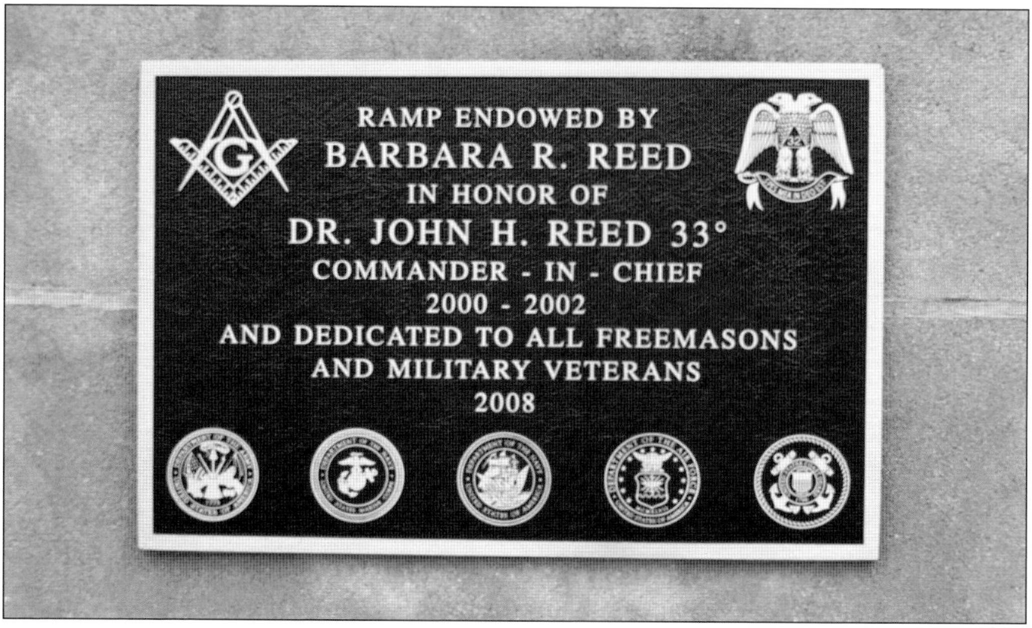

Adding a Handicap Ramp. In 2008, recognizing a need for better ingress and egress to the Cathedral, Barbara R. Reed donated funds for a ramp in honor of her husband, Dr. John H. Reed, 33rd degree, Past Commander in Chief of the Valley of New Castle Consistory. This double ramp, which rises 6 feet from the sidewalk level and spans almost 150 feet, allows the elderly, the disabled, and those carrying or wheeling packages easy access into the building.

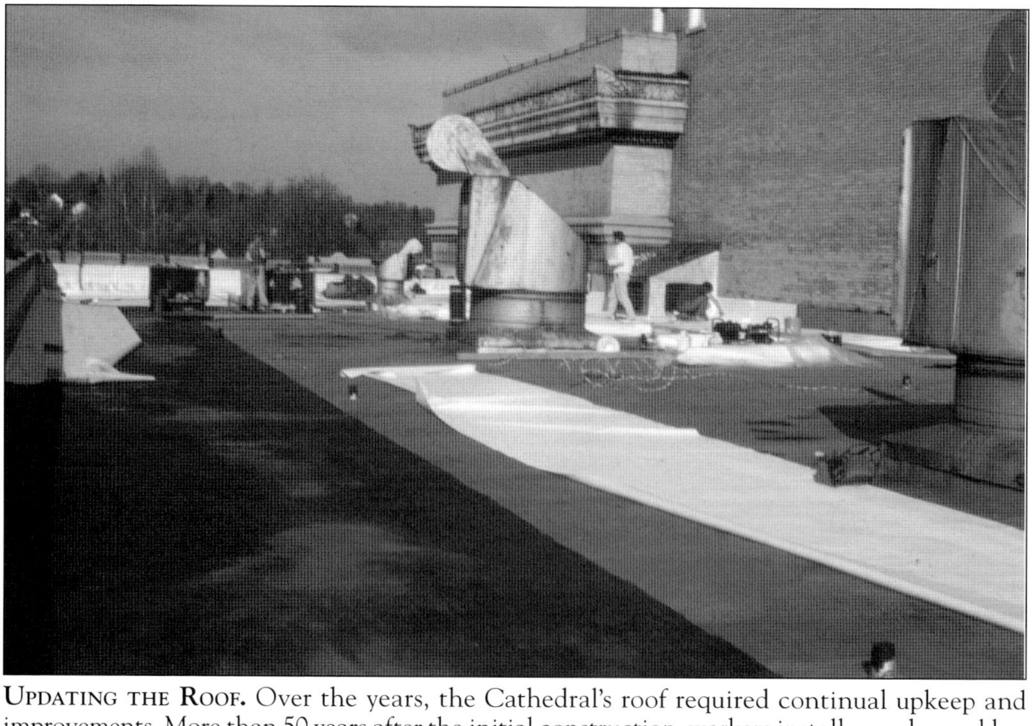

Updating the Roof. Over the years, the Cathedral's roof required continual upkeep and improvements. More than 50 years after the initial construction, workers install a modern rubber membrane to protect the building for decades to come.

Two

Laying the Cornerstone

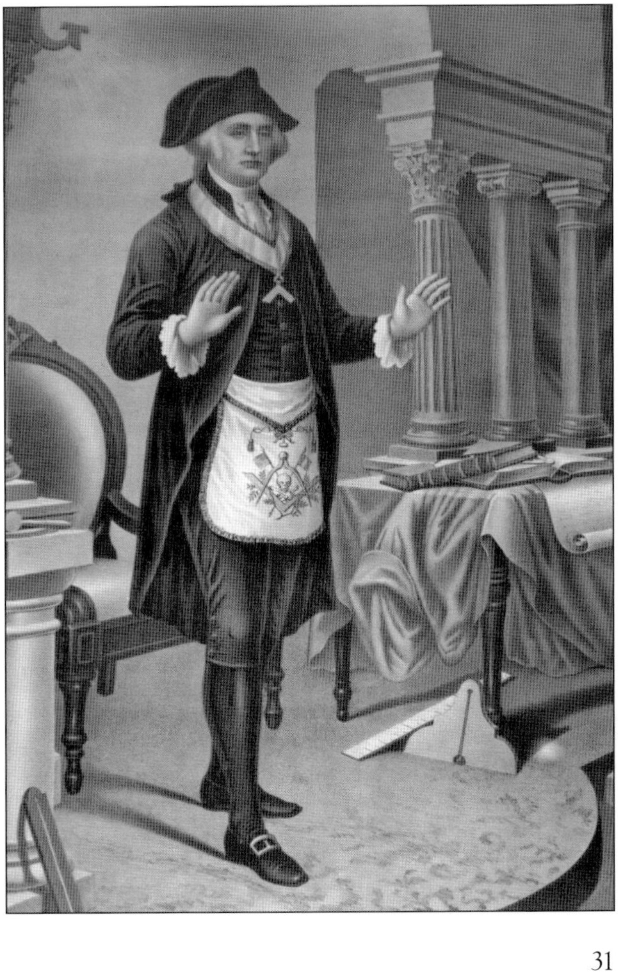

Washington Photo from Library. Pres. George Washington, one of the country's most famous Masons, proudly wears his apron and regalia. Other famous Masons of the revolutionary period included Benjamin Franklin, Paul Revere, John Hancock, and the Marquis de Lafayette.

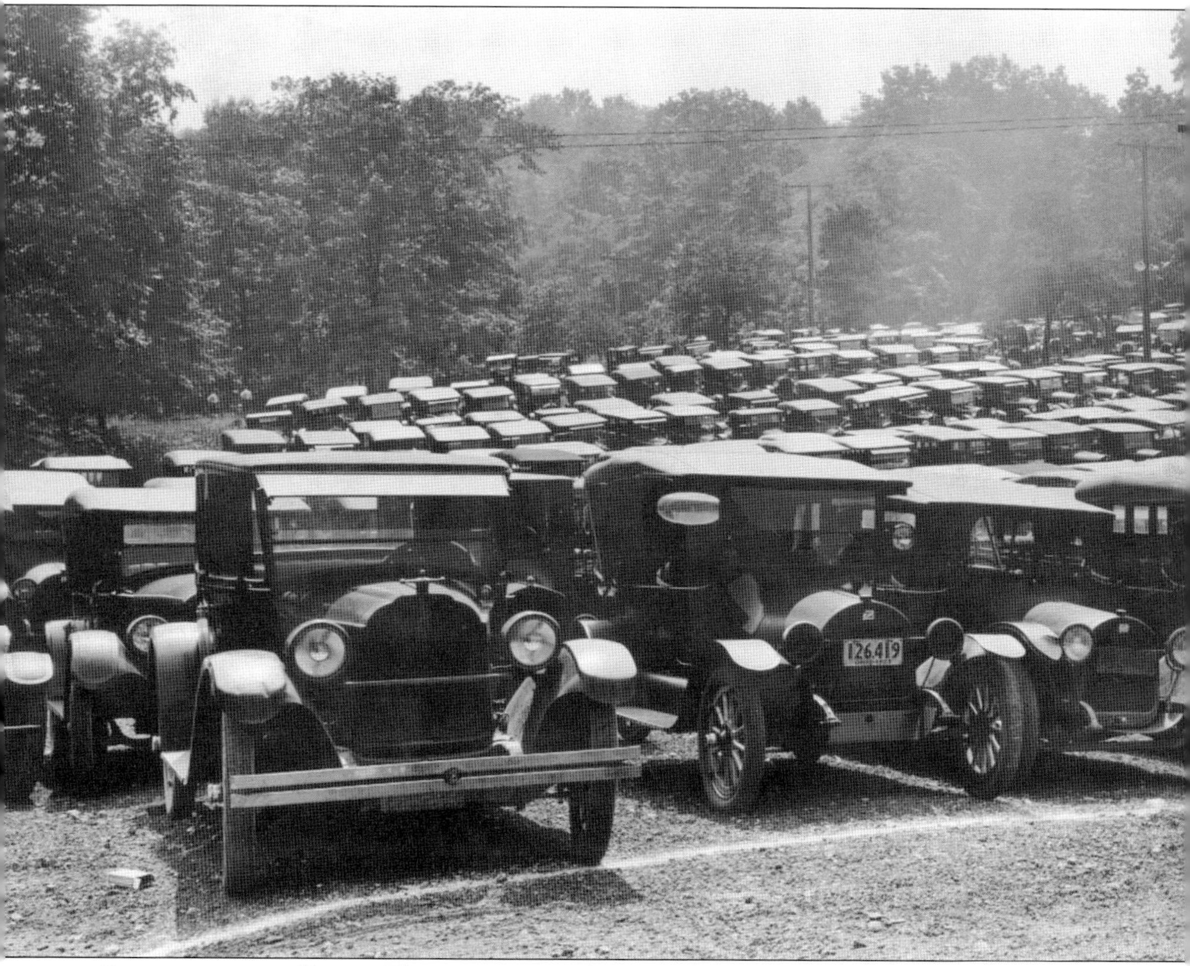

Parking for the Event. The massive attendance at the cornerstone laying ceremony held on Wednesday, June 10, 1925, forced hundreds of parkers to jam Boyles Field, four blocks north of the Cathedral on Highland Avenue. This photograph depicts the cars parked several miles away

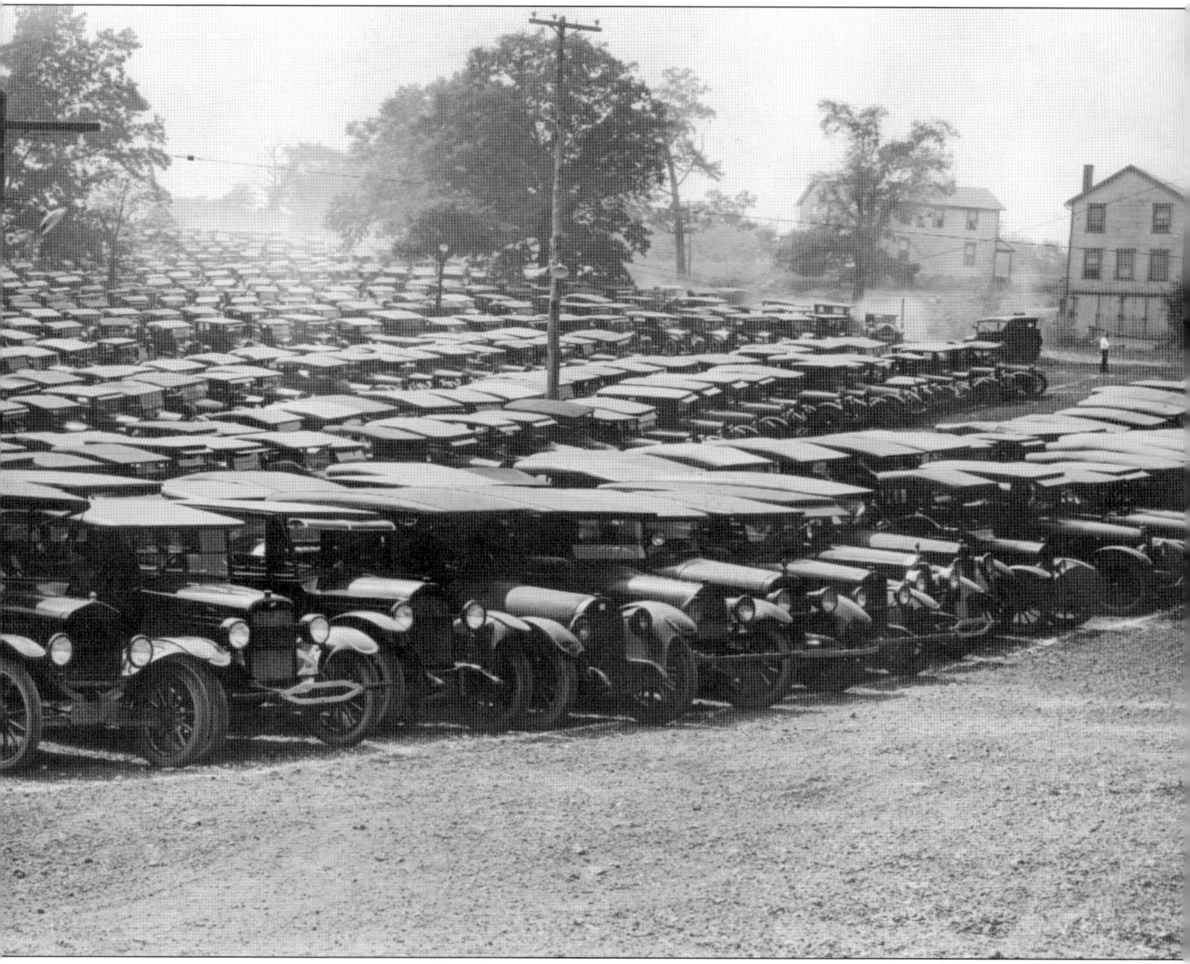

at Cascade Park. Attendees climbed aboard trolleys bound for downtown New Castle and the parade preceding the actual cornerstone laying ceremony. A festive event at the park in the late afternoon followed the Masonic ritual.

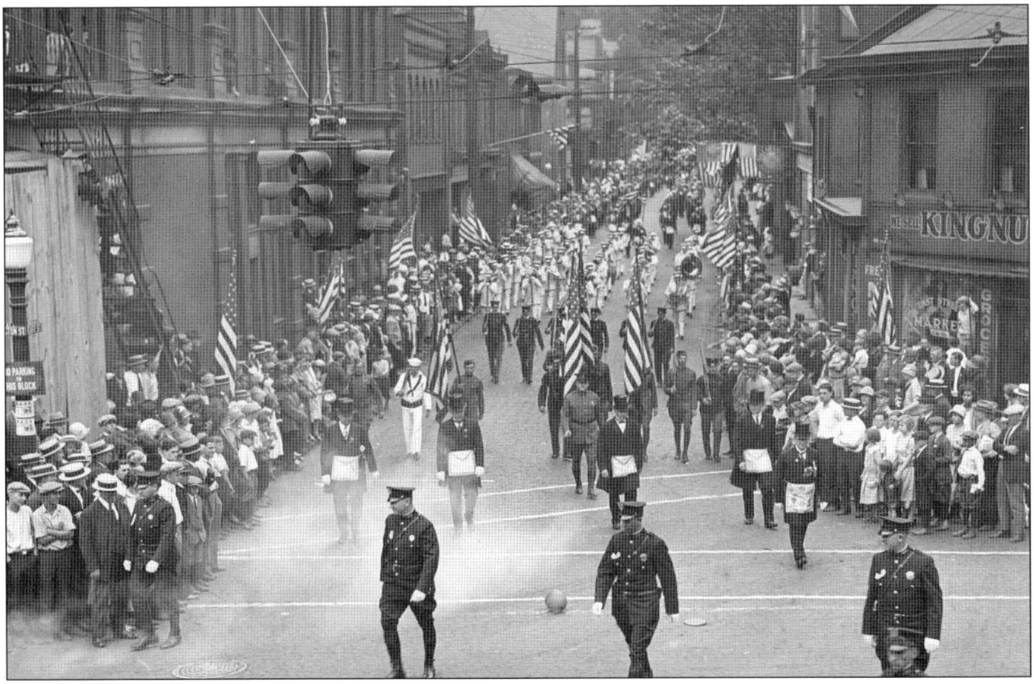

CELEBRATORY PARADE. Prior to the cornerstone ceremony, thousands of citizens lined Washington Street and the Diamond to watch the parade before it marched up the hill to the Cathedral on Lincoln Avenue. New Castle police led the way followed by grand marshal Walter S. Reynolds and the other Pennsylvania Grand Lodge officers. Mayor William H. Dunlap urged all stores to close from 11:00 a.m. until 1:00 p.m. to allow employees the opportunity to enjoy the parade and festivities.

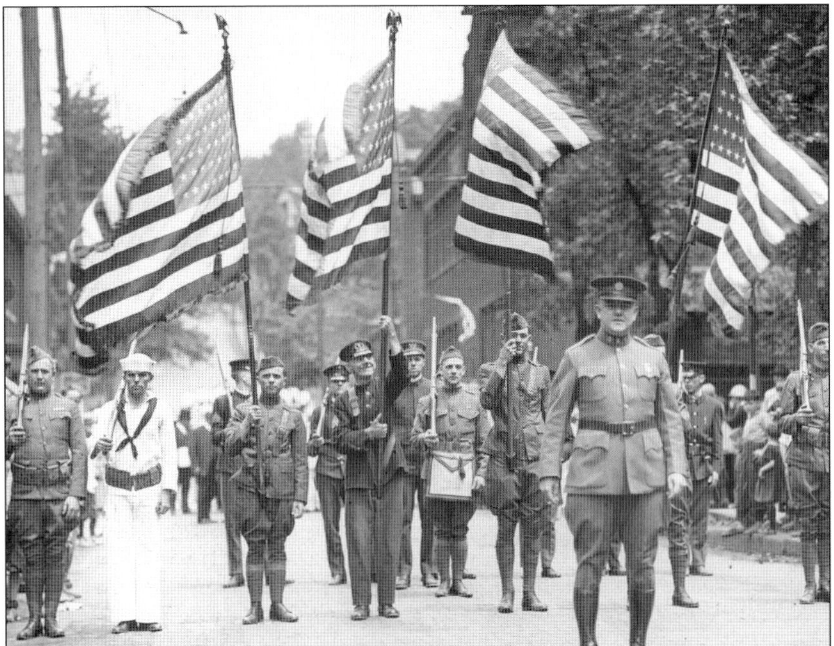

CLOSE-UP OF PARADE. World War I veterans in uniform form an honor guard for the American Flag on East Street prior to turning onto Washington Street. One soldier proudly wears his Masonic apron.

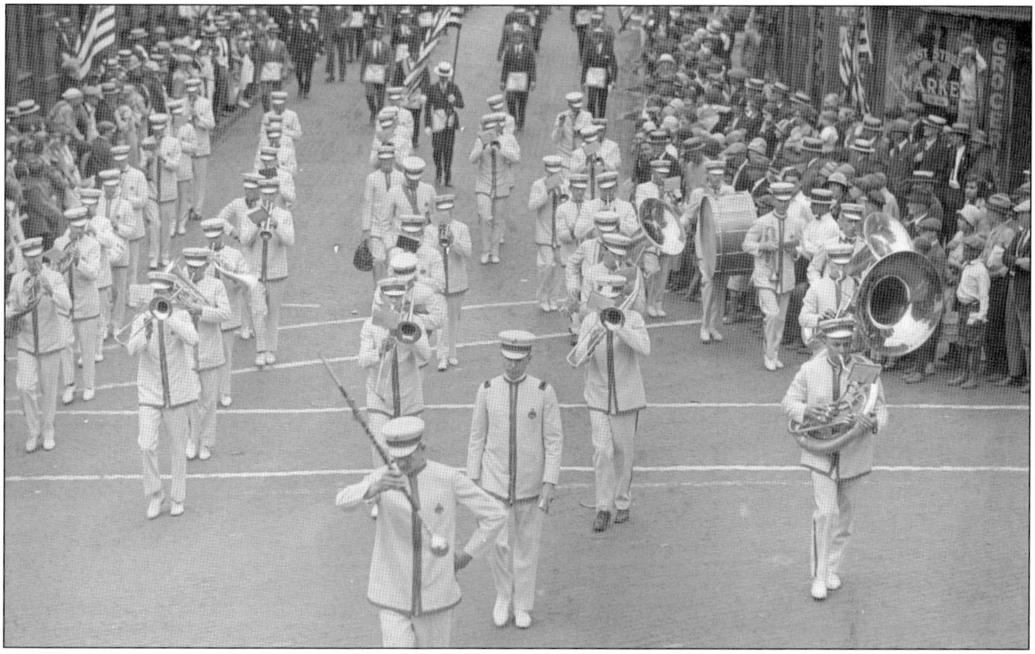

THE PARADE CONTINUES. The Scottish Rite marching band leads a group of Masons. Note the trolley tracks in the foreground and the interesting array of hats on the well-dressed male and female parade spectators.

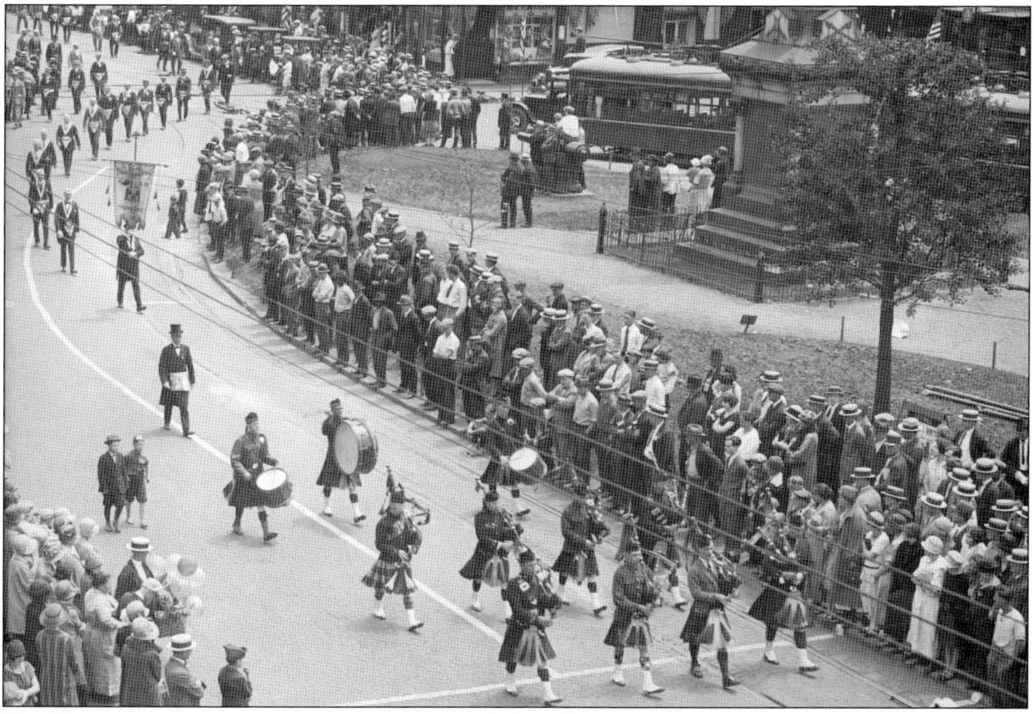

THE DIAMOND ON WASHINGTON STREET. A "kiltie" band marches toward Jefferson Street. The spectacle enthralls the children decked out in knickers. A contingent of Masons from the Valley of New Castle follow. Trolleys sit parked to the right of the Statue of the Unknown Solder.

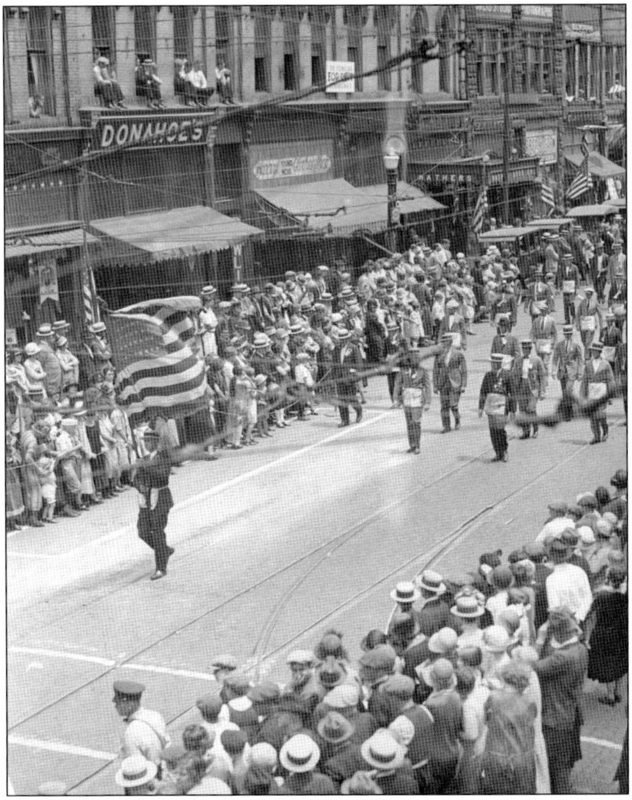

A Total of 3,000 Marching Masons. Jubilant marchers proceed up the street toward the Cathedral and the laying of the cornerstone ceremony. Spectators line the second-floor windows of downtown retail stores to gain a bird's-eye view.

Candidates in Parade. A contingent of Consistory candidates in white shawls and straw hats proudly make their way toward Lincoln Avenue for the formal cornerstone ceremony.

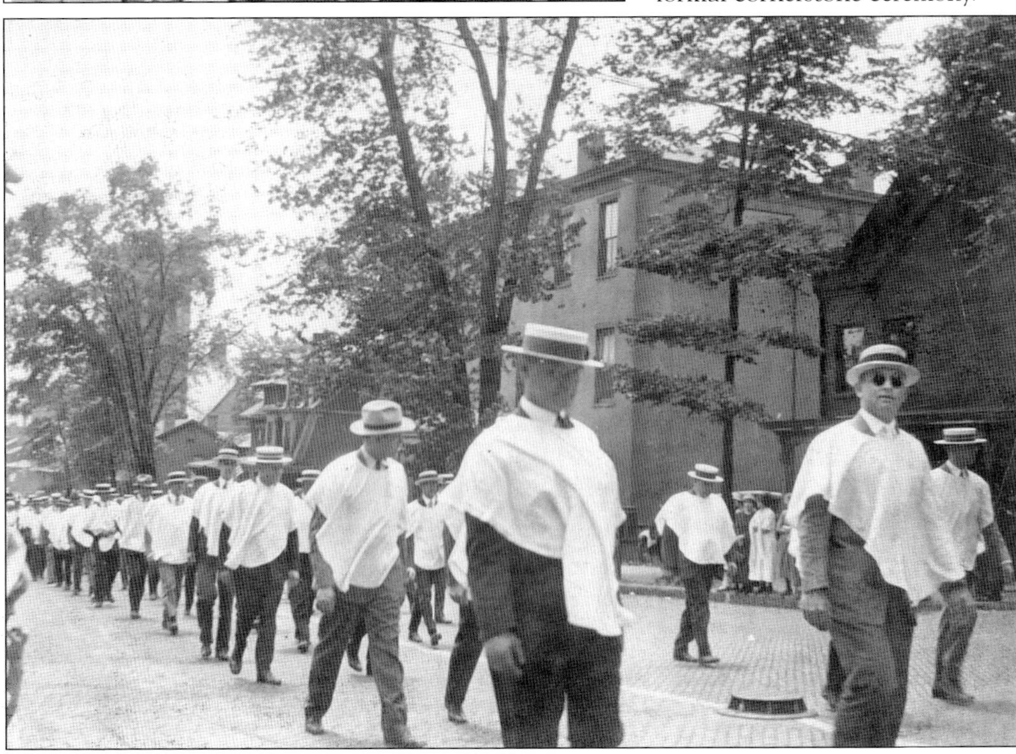

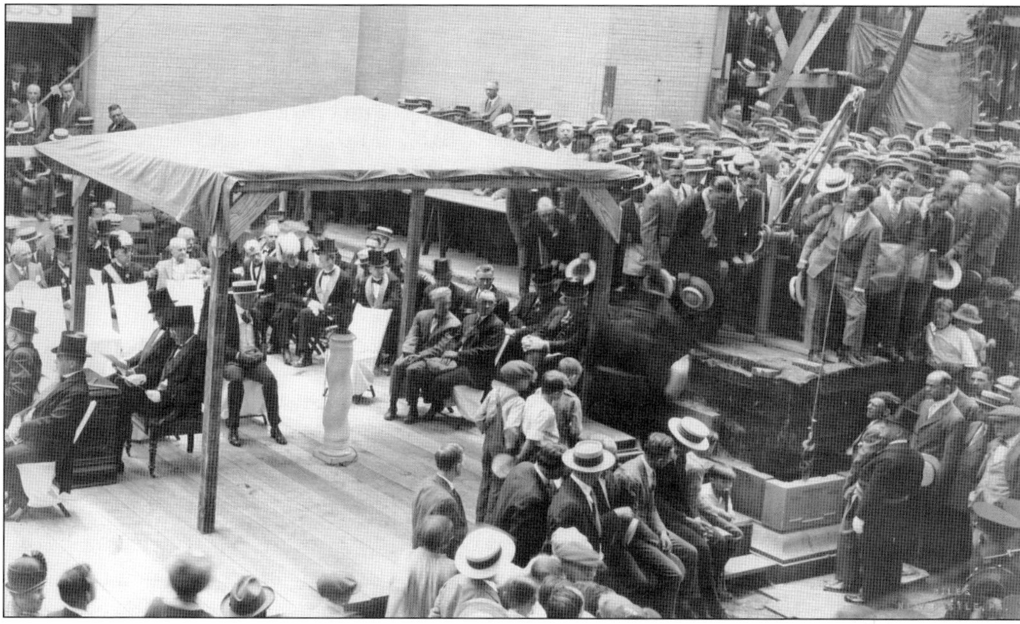

Spectators Assemble on Lincoln Avenue. Prior to the cornerstone laying, thousands of well-dressed members and citizens cram the central porch and the area around it on a clear June day to get a close-up view of the ceremony about to take place.

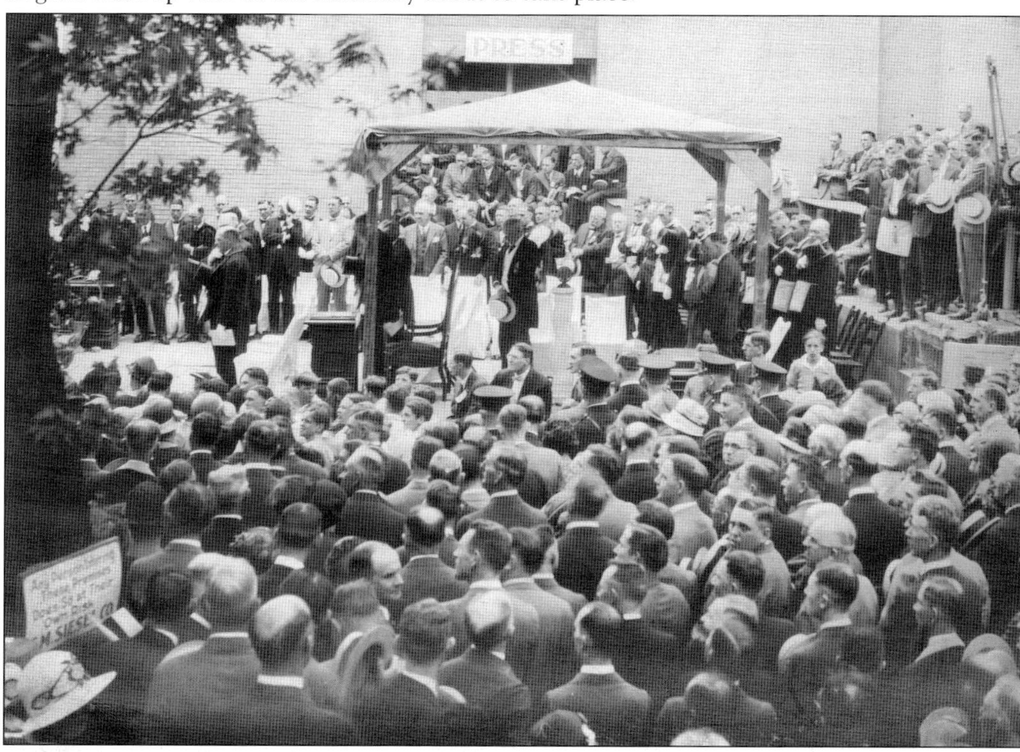

Dignitaries Fill the Podium and Front Rows. Note the cautionary sign that reads, "Any person entering these premises does so at their own risk." Work on the interior of the Cathedral had yet to be completed. Craftsmen would return to the job site Thursday morning, June 11, 1925.

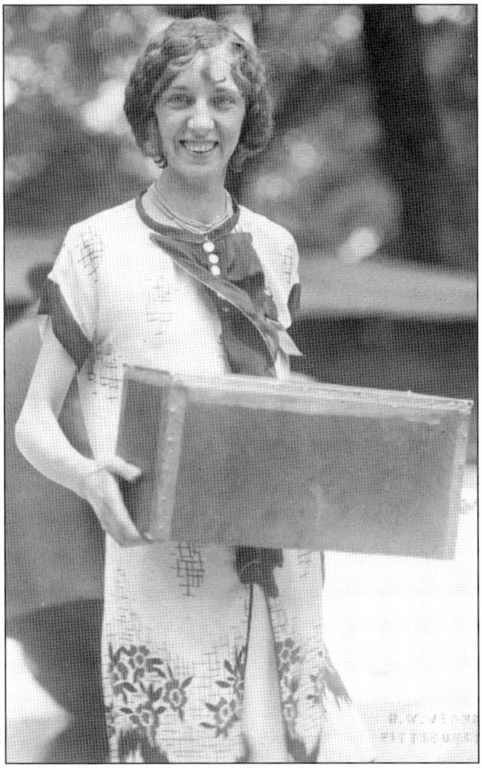

Time Capsule Copper Box. Prior to the placement of the cornerstone, Mrs. W.F. Jackson carried a time capsule containing earth from the Garden of Gethsemane, the Holy Bible, histories of Lodge Nos. 243, 642, 727, and 433, copies of the *New Castle News*, 1925 coins totaling $1.73, a chapter penny, a photograph of John S. Wallace, a program from the Scottish Rite reunion of 1925, work site and progress pictures, two $1,000 Cathedral bonds, the names of the Grand Lodge officers participating in the day's program, and the member names of the Cathedral Association. The Grand Lodge of Pennsylvania decreed that the time capsule should be opened on or around June 10, 2025.

Placement of Cornerstone. The Right Worshipful Grand Master of the Grand Lodge of Pennsylvania, Samuel M. Goodyear, positions the cornerstone according to proper ritual. The Right Worshipful Senior Warden stands attentively with a 24-inch scale to insure exact alignment. A trowel and mortar stand ready for sealing the stone into the northeast corner of the building at the right moment in time.

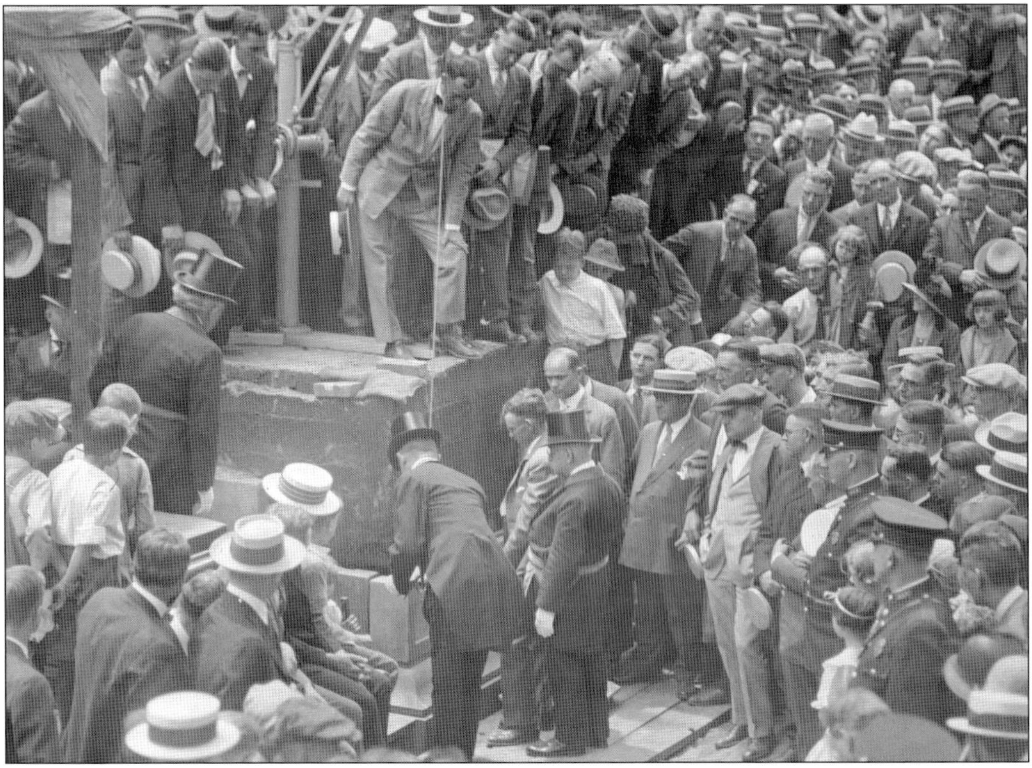

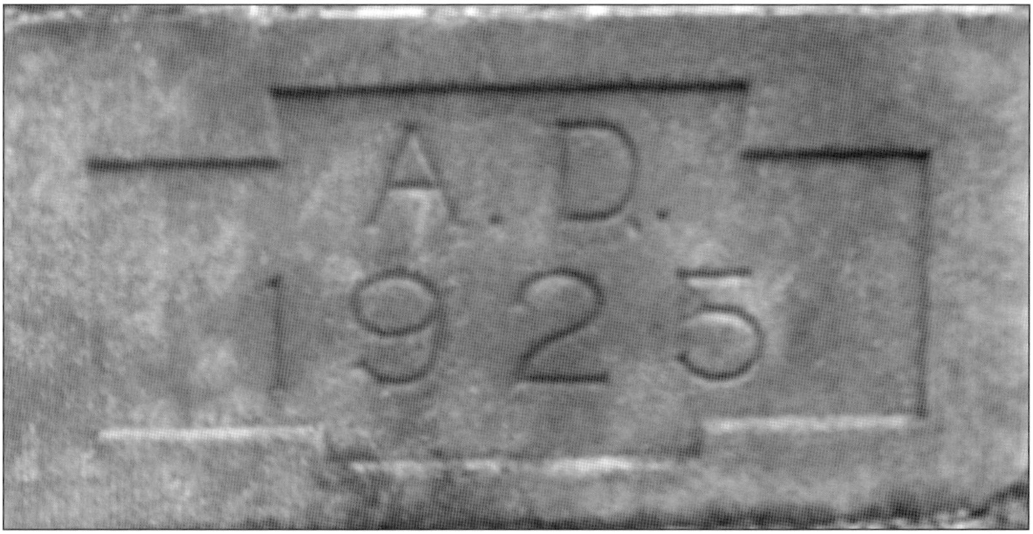

The Cornerstone. The keynote speaker, the Illustrious James Isaac Buchanan, explained thusly: "A cornerstone is the foundation on which the entire structure is supposed to rest. It was considered by the ancient operative Masons as the most important stone in the edifice, hence the cornerstones were laid by them with impressive ceremonies. The symbolism is full of significance as to its form, situation, its permanence, and its consecration. As to form, it must be perfectly square on its surface and in its solid content a cube. The square is a symbol of morality and the cube of truth in its situation. It lies between the North, a place of darkness, and the East, a place of light, and symbolizes Masonic progress from dark to light, from ignorance to knowledge."

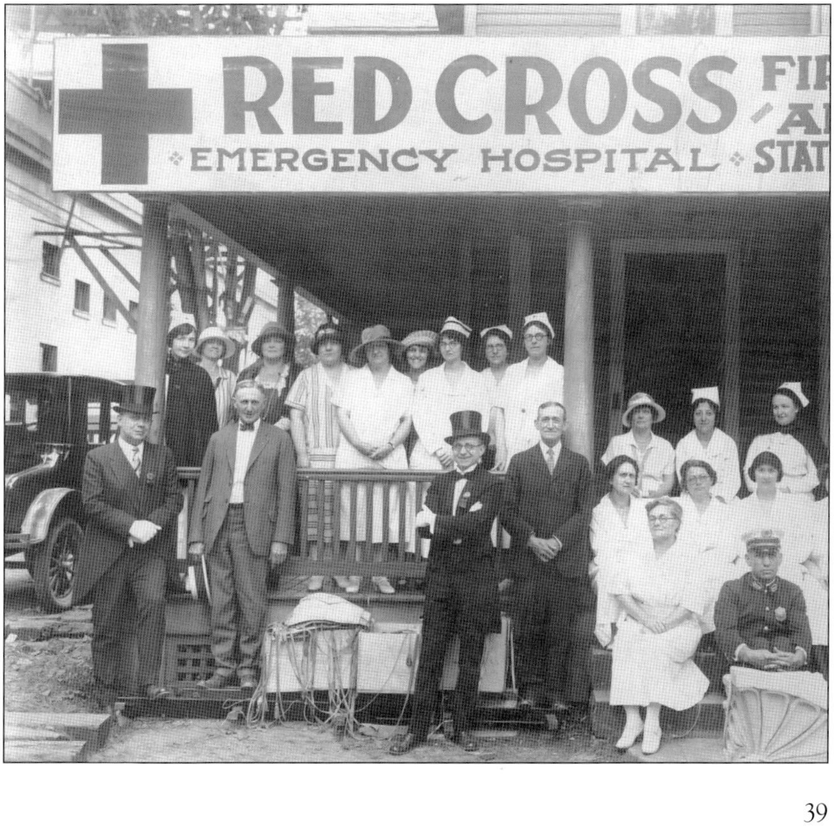

Safety and Security. The Red Cross set up a first-aid station at the neighboring Miller home on the west side of the Cathedral staffed by nurses and doctors. Local police insured order and crowd control. A Masonic official in a top hat stands ready to collect aprons at the conclusion of the ceremony.

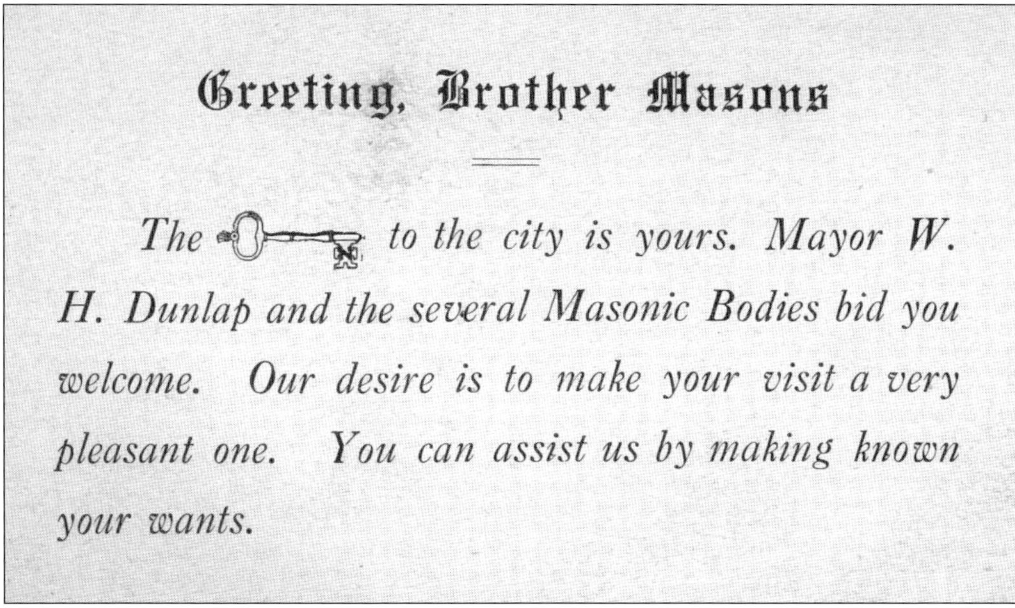

Greeting, Brother Masons

The [key] to the city is yours. Mayor W. H. Dunlap and the several Masonic Bodies bid you welcome. Our desire is to make your visit a very pleasant one. You can assist us by making known your wants.

KEY TO THE CITY. Local dignitaries treated visitors to the Lawrence County community with the upmost respect. New Castle mayor W.H. Dunlap provided a key to the city to out-of-town Masonic attendees.

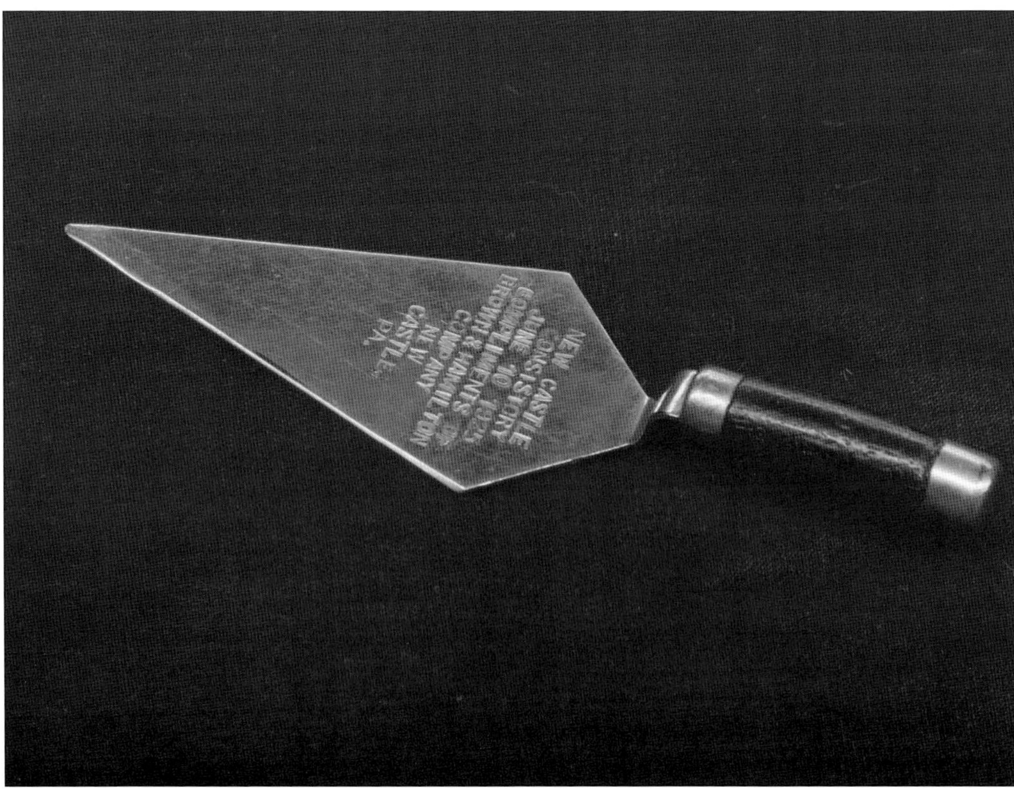

ENGRAVED COMMEMORATIVE TROWEL. Visiting dignitaries received this handsome gift manufactured by Brown & Hamilton as a souvenir of the day's celebration.

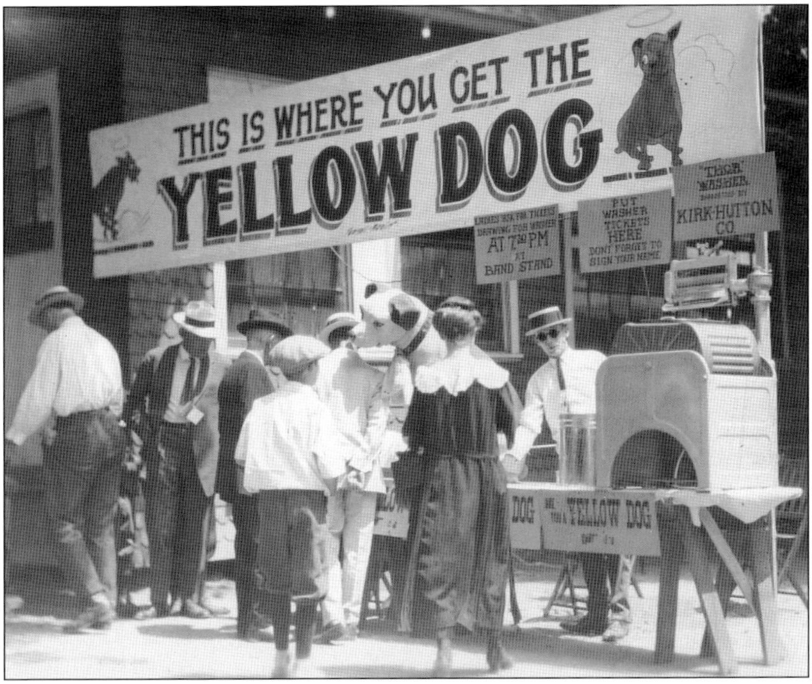

Registration for Gifts and Prizes. Residents line up for a free drawing on a chance to win a washing machine donated by the Kirk Hutton Store. A variety of food vendors did a brisk business throughout the day.

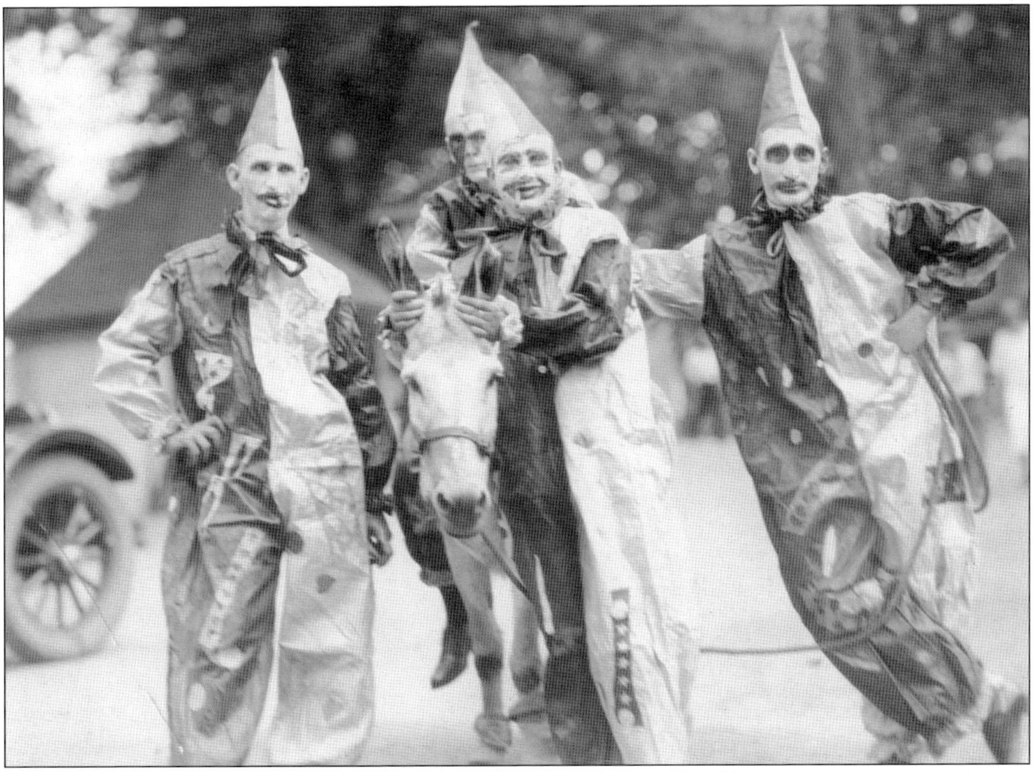

Entertainment for All Ages. Clowns added the perfect touch of levity and merriment for the populace. Children and adults alike spoke of the wide array of food and attractions for months to follow.

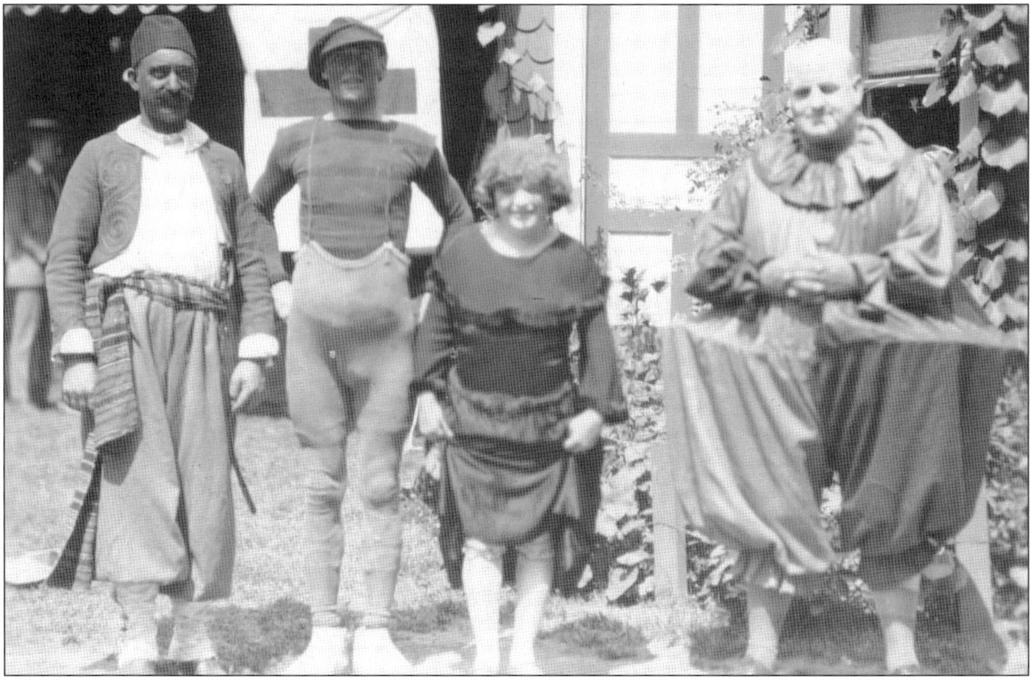

ACTORS PROVIDE HUMOR. A costumed crew poses for photographs and jokes with attendees, putting everyone into a joyous mood. Many of the performers happened to be Masons who entertained as volunteers.

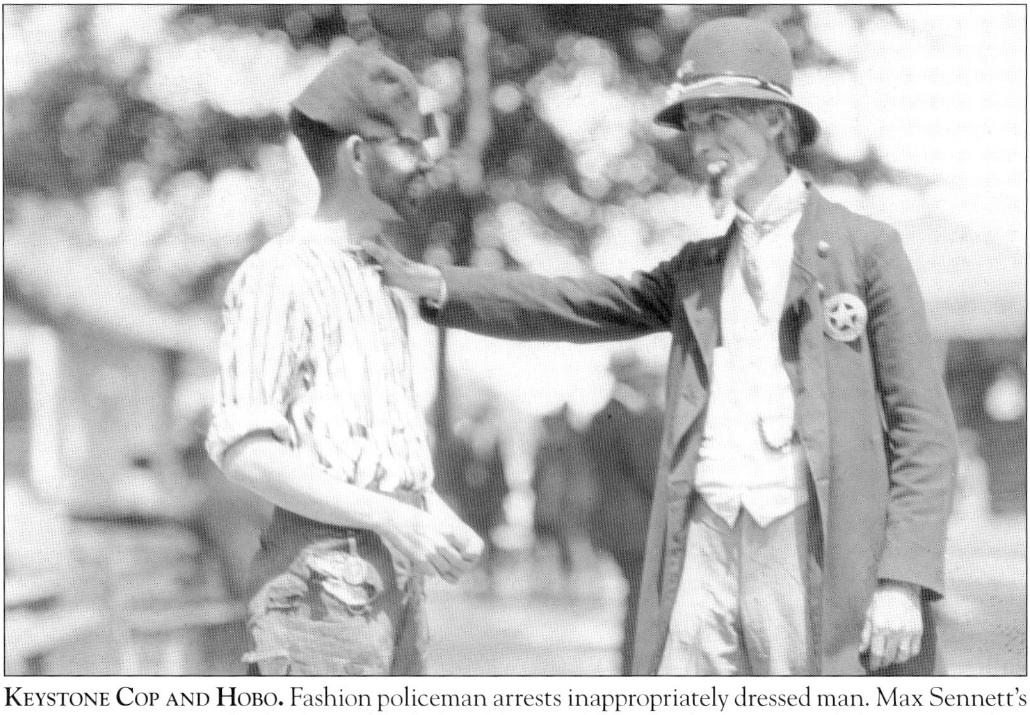

KEYSTONE COP AND HOBO. Fashion policeman arrests inappropriately dressed man. Max Sennett's *Keystone Cops* series had achieved unbelievable popularity in the 1920s, and this pair made a guaranteed hit with the audience.

Three

Valley of New Castle

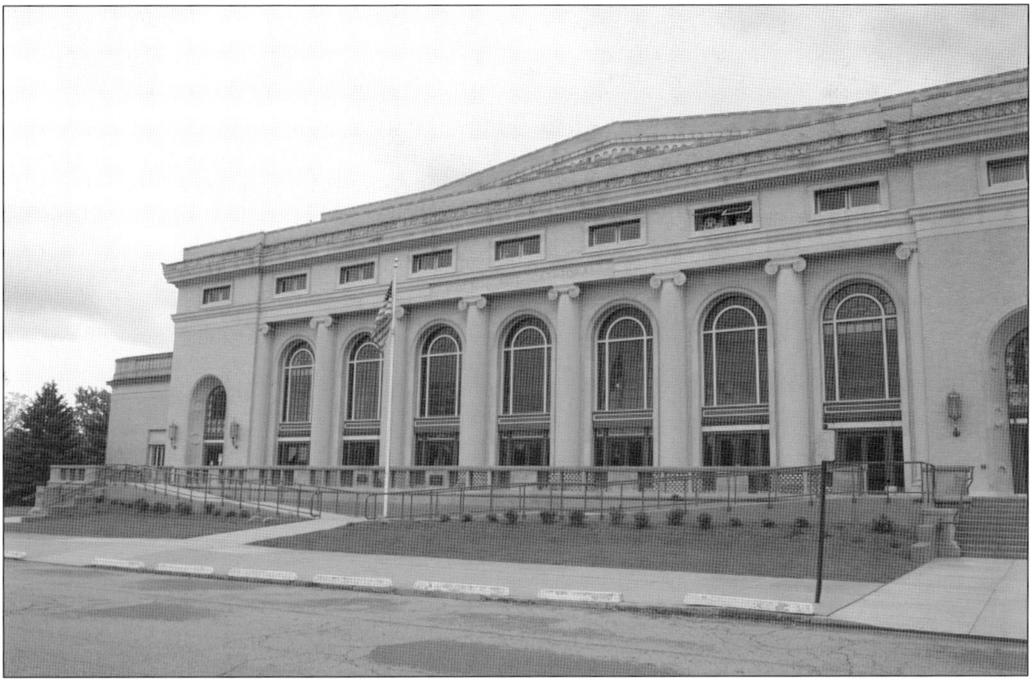

Northwest view of Cathedral. This building stands as the home for Scottish Rite Masons in a 10-county area spanning more than 65 miles from New Castle. A man's journey in Freemasonry begins with the Entered Apprentice degree and continues with the Fellowcraft and Master Mason degrees, commonly known as the first, second, and third degrees. The Ancient Accepted Scottish Rite (AASR) encompasses the next 29 degrees divided into four categories known as: Lodge of Perfection, Council of Princes of Jerusalem, Chapter of the Rose Croix, and Consistory. Upon receiving the 32nd degree, the Mason becomes a Sublime Prince of the Royal Secret, whose privilege allows him to roam in his intellect across the infinite wonders of the universe, while with his hands he performs the humble labors of earth. The Valley of New Castle serves much of Western Pennsylvania, including all of Beaver, Lawrence, Mercer, Venango, Butler, Forest, Clarion, Armstrong, Jefferson, and Indiana Counties. Since the Consistory allows for dual memberships, the brotherhood contains a few members from outside these counties.

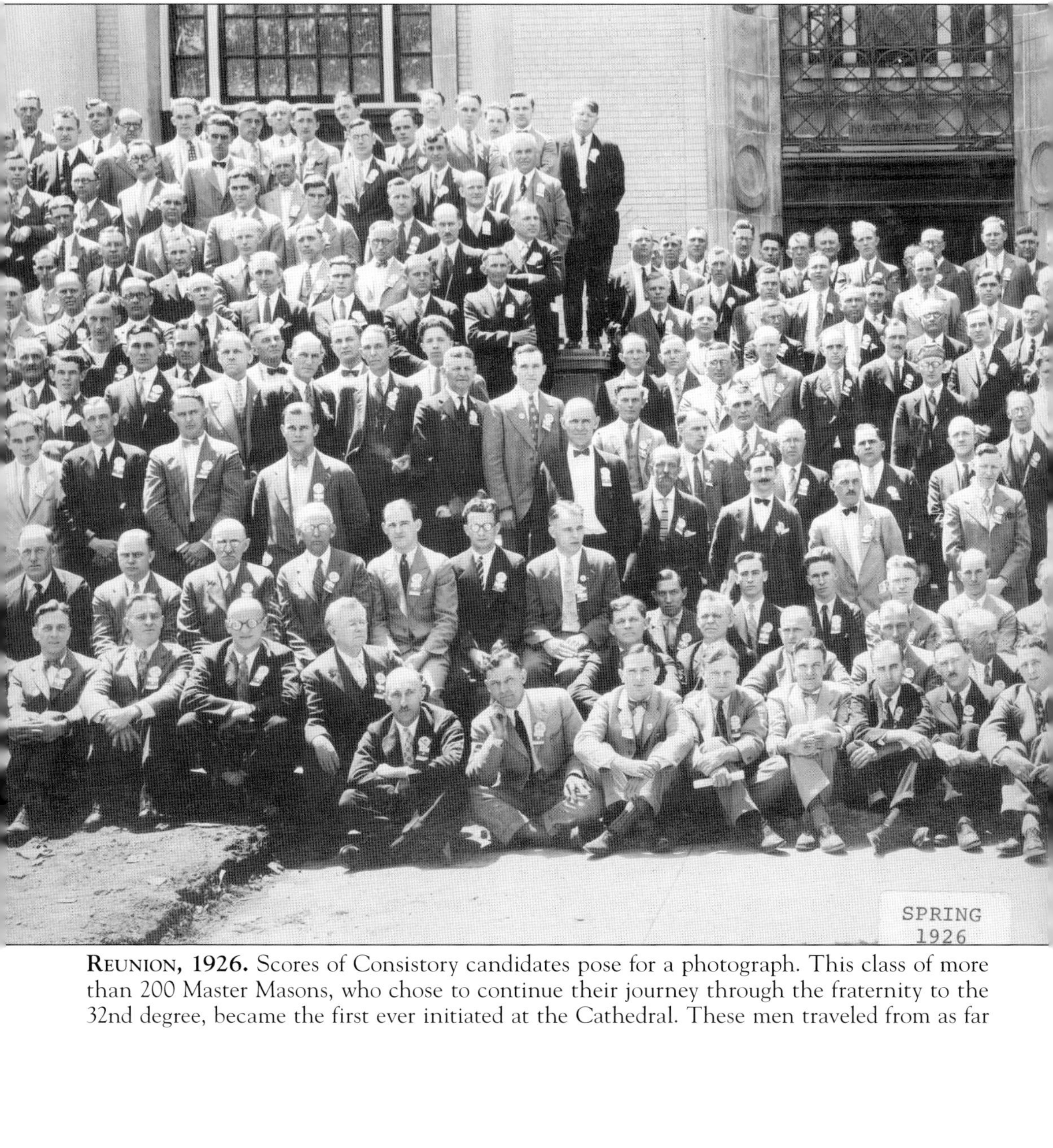

Reunion, 1926. Scores of Consistory candidates pose for a photograph. This class of more than 200 Master Masons, who chose to continue their journey through the fraternity to the 32nd degree, became the first ever initiated at the Cathedral. These men traveled from as far

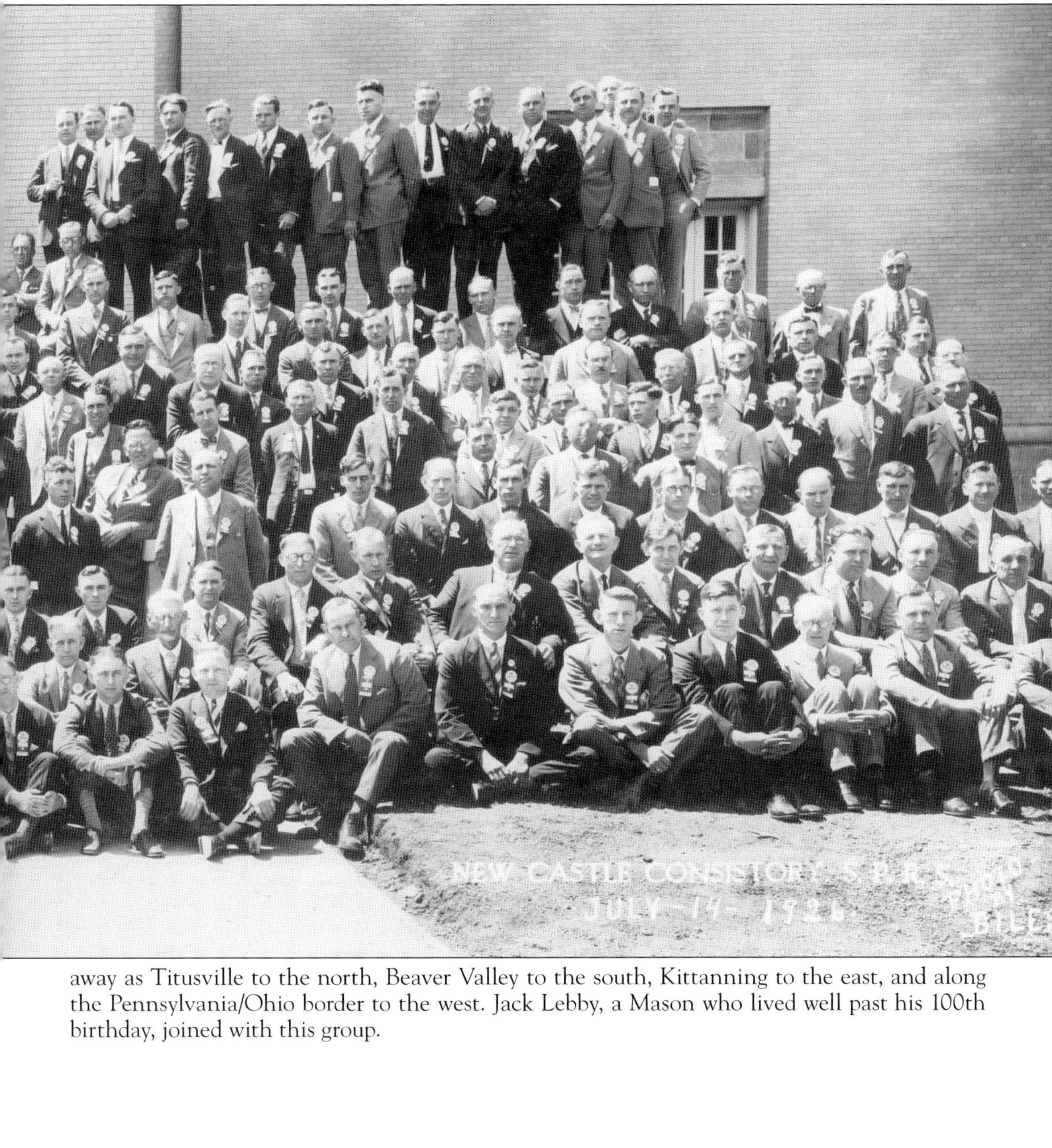

away as Titusville to the north, Beaver Valley to the south, Kittanning to the east, and along the Pennsylvania/Ohio border to the west. Jack Lebby, a Mason who lived well past his 100th birthday, joined with this group.

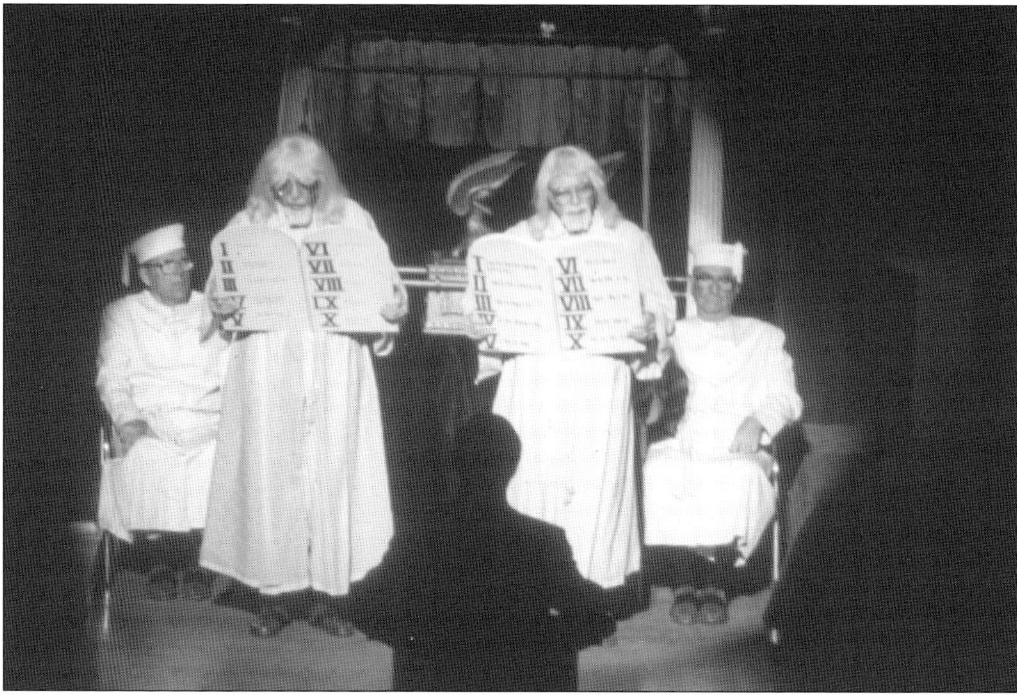

14TH DEGREE, GRAND ELECT MASON. This degree teaches us reverence to God and to His Holy name. Moses reads the Ten Commandments in both Hebrew and English; thus, two members portray the prophet. The scene takes place below the Sanctum Sanctorum and features King Solomon of Israel and King Hiram of Tyre, among others.

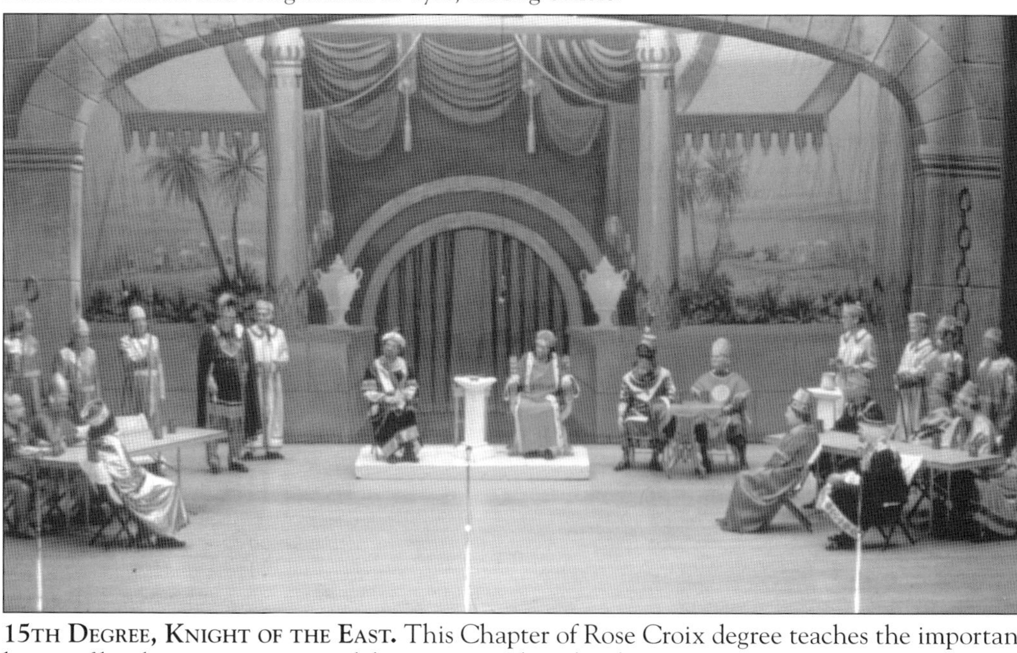

15TH DEGREE, KNIGHT OF THE EAST. This Chapter of Rose Croix degree teaches the important lesson of loyalty to conviction and devotion to right. The character King Cyrus of Persia (an older cousin and predecessor to Darius) and his nobles confer with the Knights of the East regarding the rebuilding of the ancient Temple of Solomon, destroyed in 422 BC.

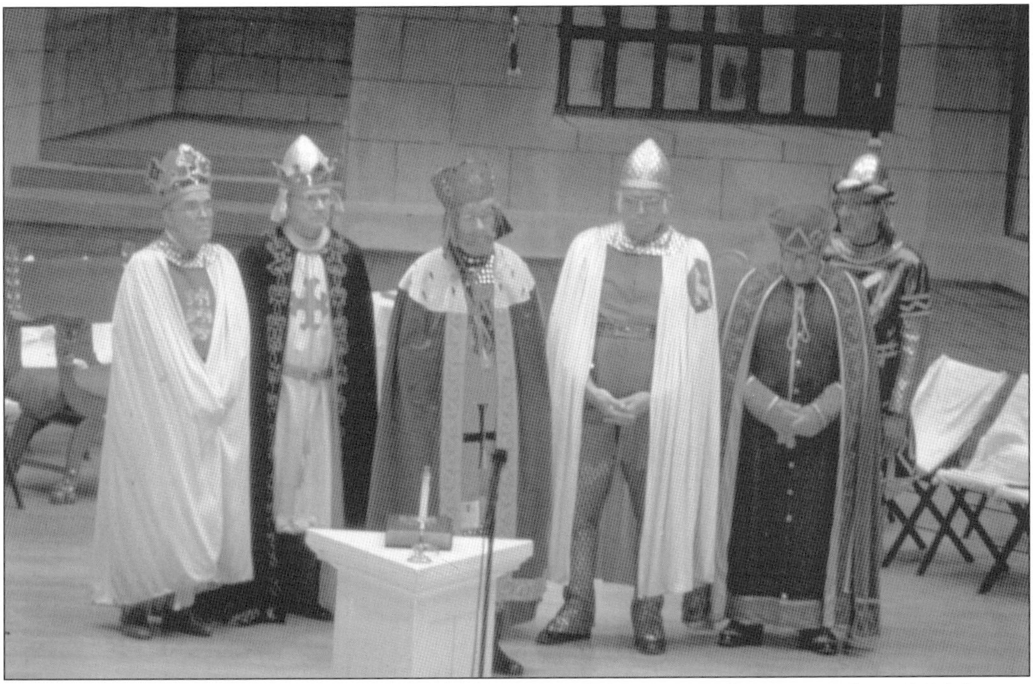

16TH DEGREE, PRINCE OF JERUSALEM. This historical degree emphasizes truth and fidelity to duty. King Darius of Persia grants an audience to the Knights of the East at his throne room in Babylon.

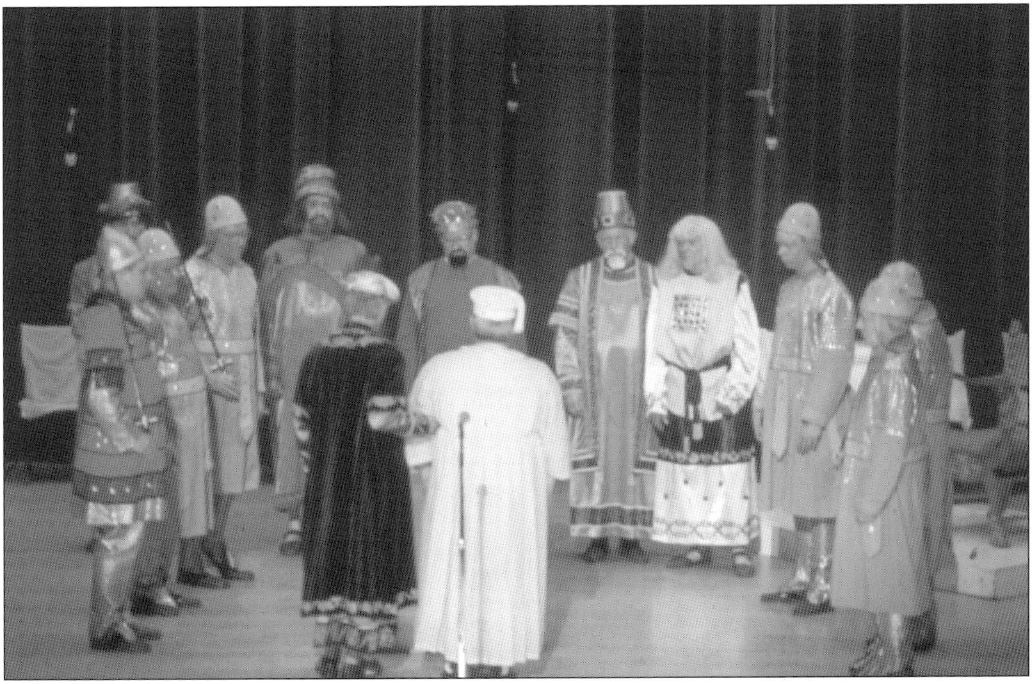

18TH DEGREE, KNIGHT OF THE ROSE CROIX. This degree affirms the broad principles of universality and tolerance. Love one another. The scene shows a Neophyte, Masonic officers, a Captain of the Guard, and several assisting knights.

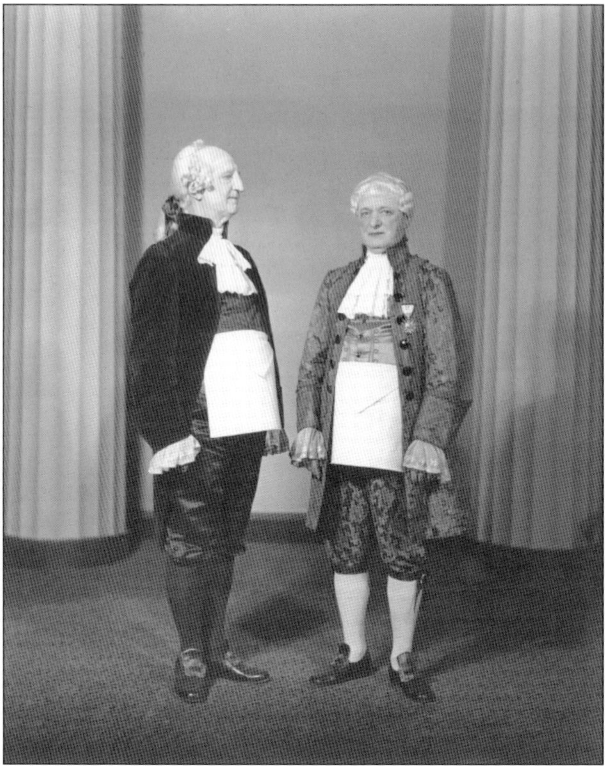

20TH DEGREE, MASTER AD VITAM OF THE COUNCIL OF PRINCES OF JERUSALEM. This degree, highlighting patriotism, demonstrates the Masonic condemnation of all who conspire against the security of the nation and the happiness of our people. George Washington and the Marquis de Lafayette (played by Samuel Hadley and Reed Megown, both 33rd degrees) hold accountable Benedict Arnold and British colonel Gilbert Beltower for their treachery.

ADDITIONAL CAST MEMBERS OF THE 20TH DEGREE. Gen. Nathanial Green of Rhode Island, Gen. Henry Knox of Massachusetts, and Col. Alexander Hamilton of New York also played key patriotic roles in the allegorical tale found in the second degree of the so-called chivalric and philosophical grades.

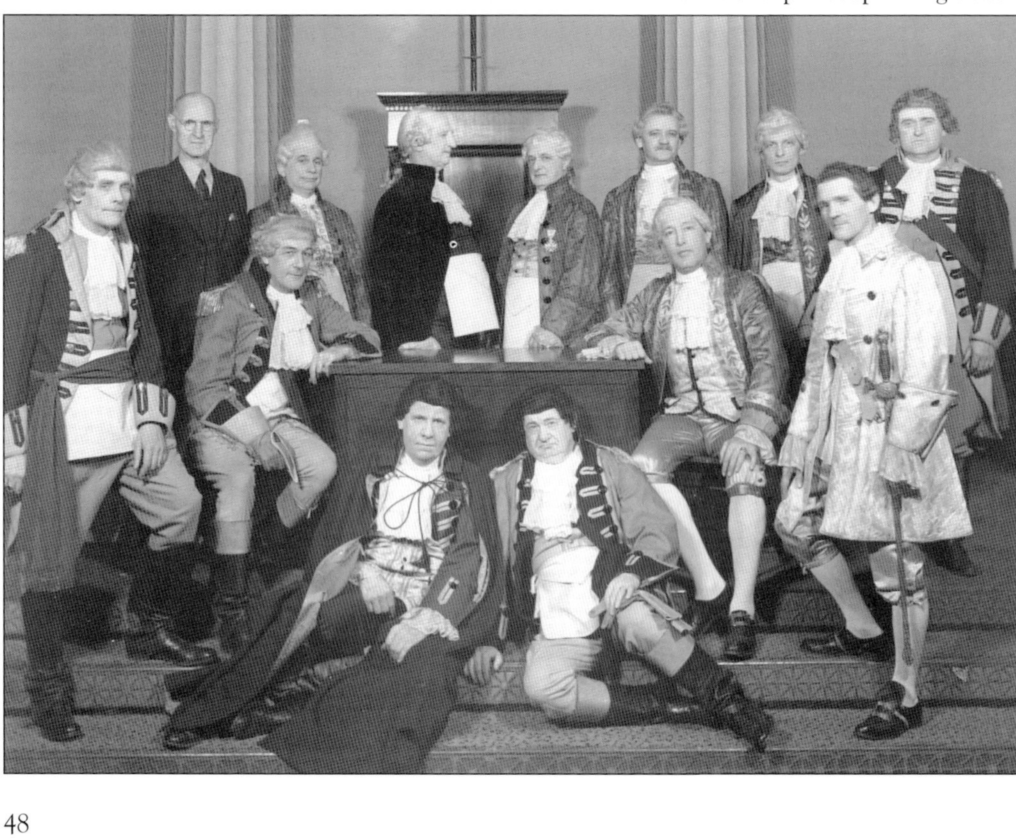

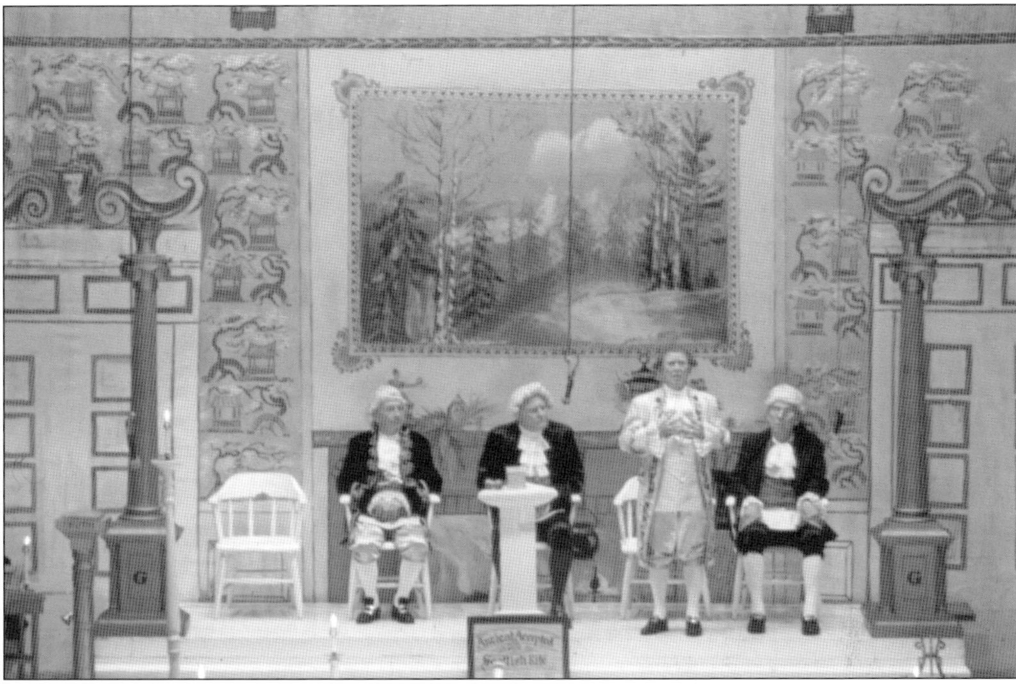

Later Presentation of the 20th Degree. This exemplification took place on the stage of the auditorium rather than in the east lodge room as shown on the facing page, allowing for sophisticated backdrops, props, and scenery. Sam Mullin (seated behind podium) portrayed Gen. George Washington, and Louis Berkheimer (standing) played Marquis de Lafayette.

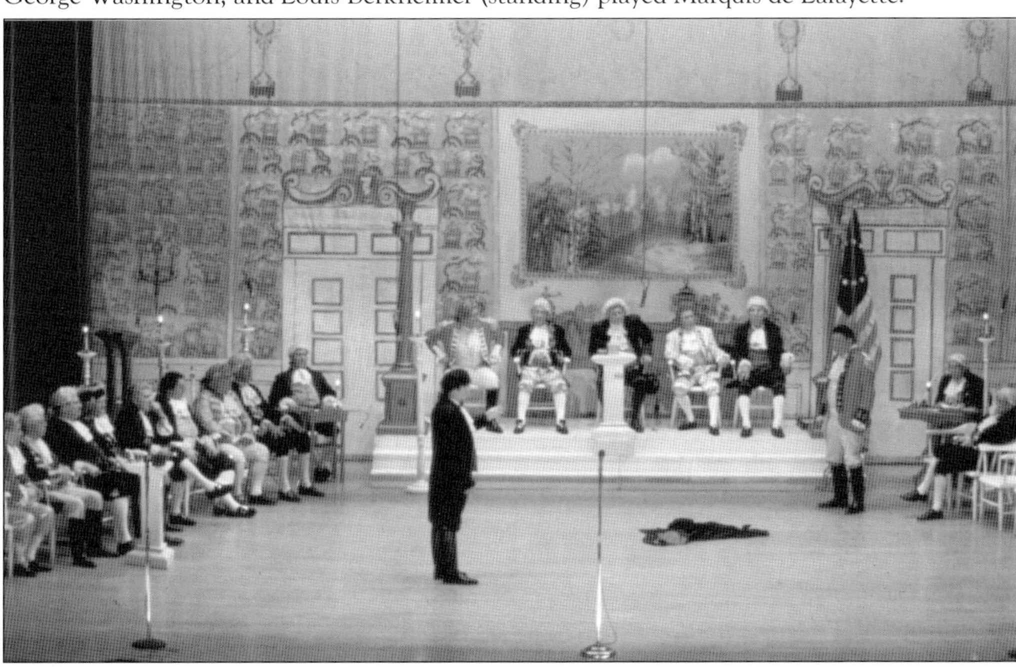

The Accused Stand Unmasked. Masonic principles and leadership are subjected to a crucial test. American patriots reveal the treachery of Benedict Arnold (standing and dressed in black) and Col. Gilbert Beltower (standing in front of flag) for all to see.

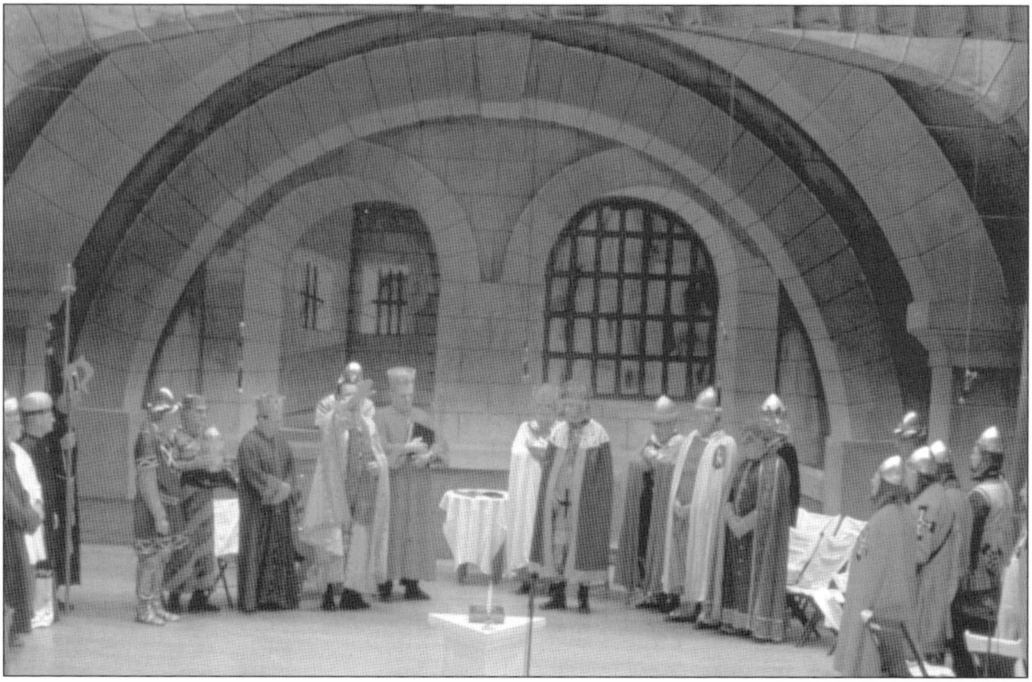

19th Degree, Grand Pontiff. This degree proclaims this spiritual unity of all who believe in God and cherish the hope of immortality, no matter what religious leader they follow or creed they profess. The lesson highlights the conflict between light and darkness, good and evil, God and Satan. The cast includes the Grand Pontiff, his counselors, and knights.

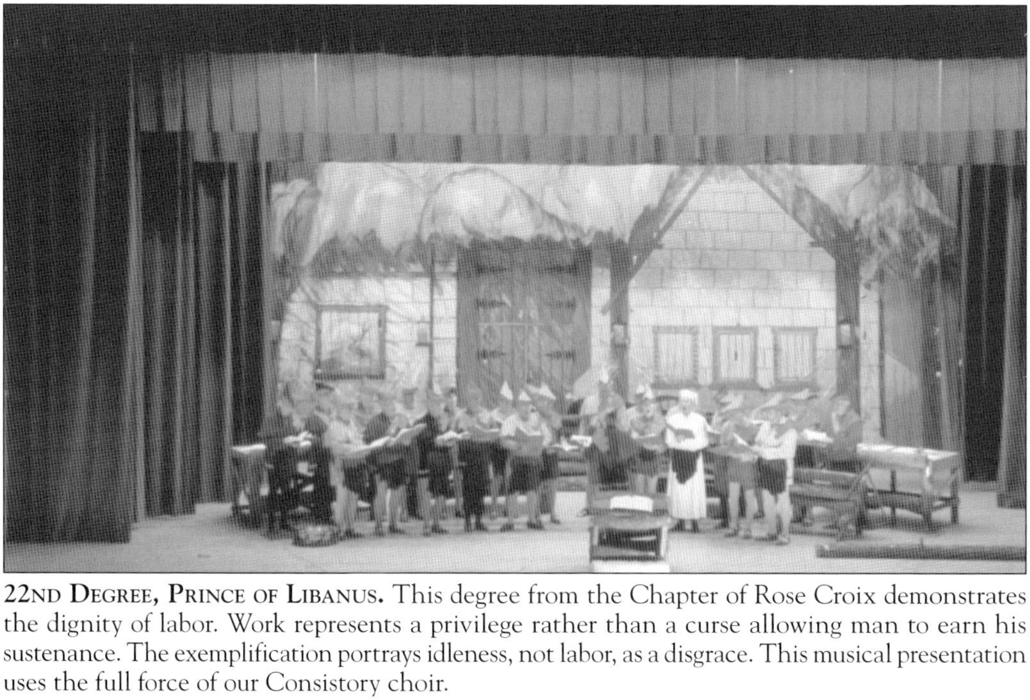

22nd Degree, Prince of Libanus. This degree from the Chapter of Rose Croix demonstrates the dignity of labor. Work represents a privilege rather than a curse allowing man to earn his sustenance. The exemplification portrays idleness, not labor, as a disgrace. This musical presentation uses the full force of our Consistory choir.

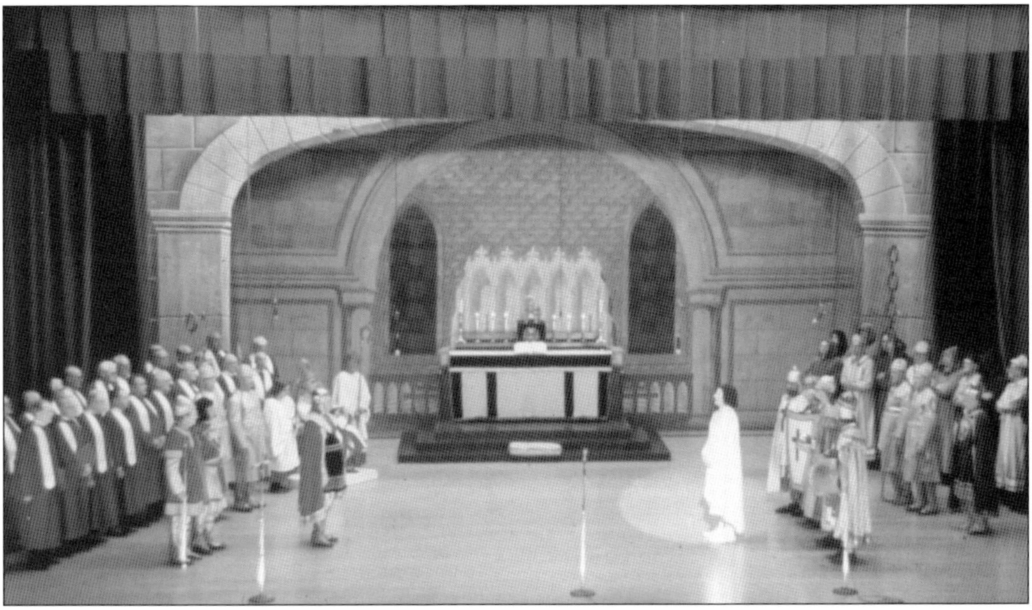

32nd Degree, Sublime Prince of the Royal Secret. This degree represents the 15th and final of the chivalric and philosophical grades. Constans stands in as the exemplar for all the initiates in the class as he remains firm in his beliefs regardless of the suffering he will surely face.

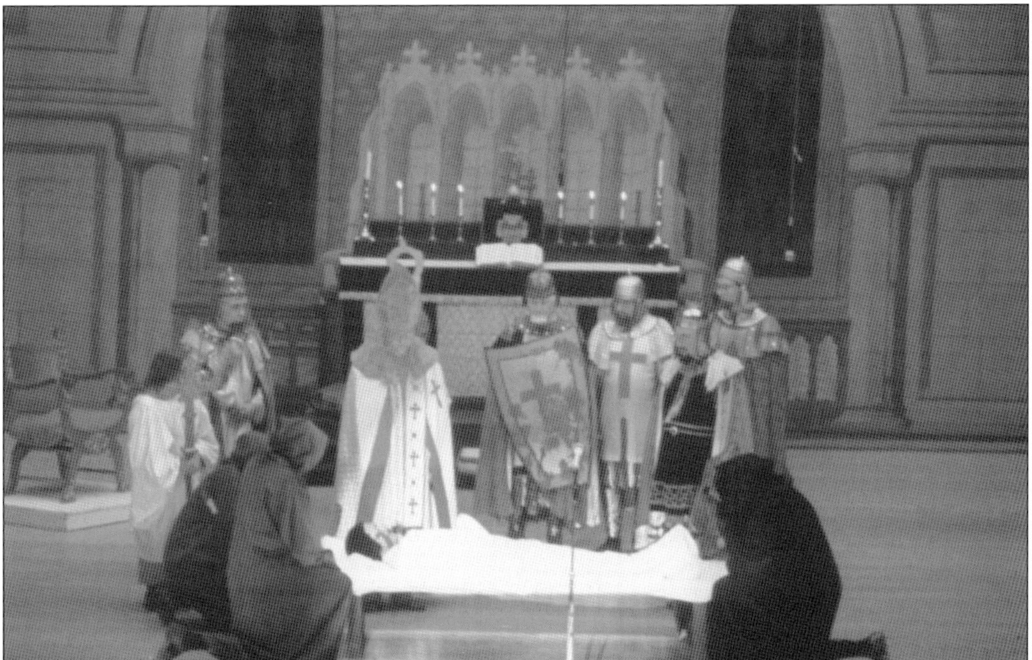

32nd Degree, The Sacrifice. As the culmination of Scottish Rite Freemasonry, this degree of Consistory celebrates the triumph of individual integrity and our obligation to serve humanity. The duty often requires sacrifice—sometimes, the supreme sacrifice.

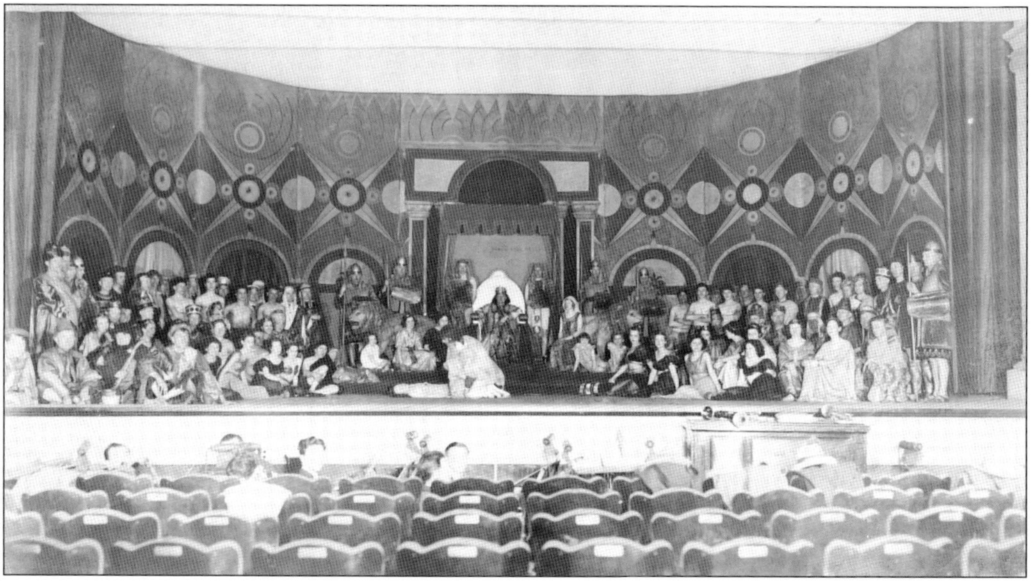

The Musical *Darius*. The Service Division from the New Castle Consistory staged *Darius* on March 29, 1935. The spectacular production featured a cast of 100, a full-symphony orchestra, and a chorus to tell the story of the powerful Persian king, who ruled from 550 to 486 BC. The Consistory welcomed the general public to the musical.

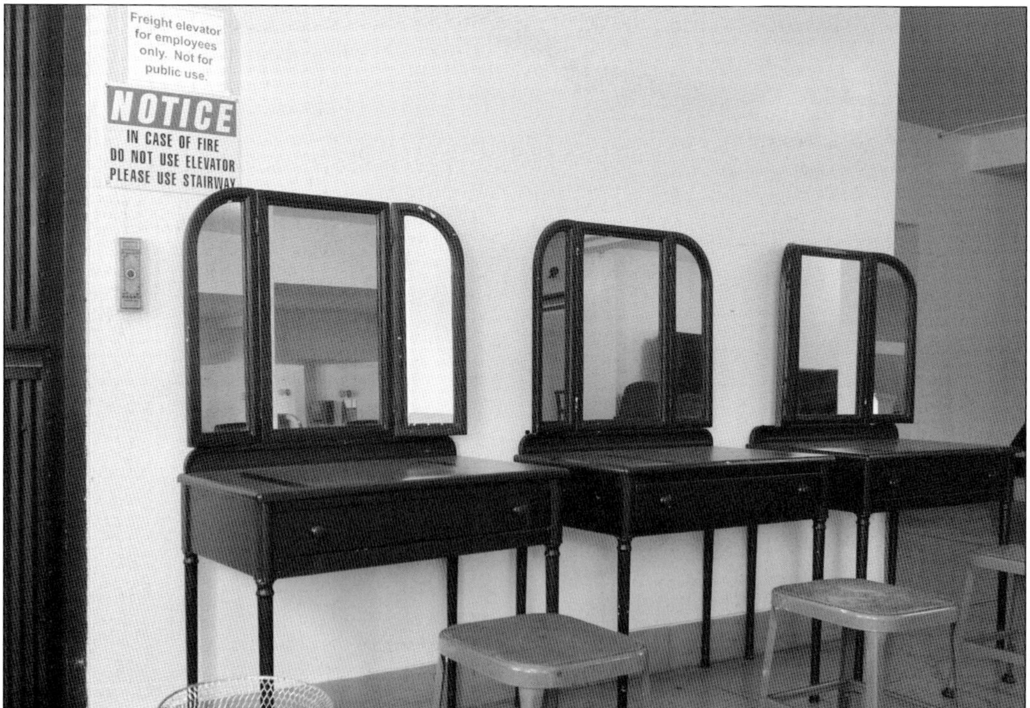

Antique Makeup Tables. A separate section of the makeup area includes a number of makeup tables with adjustable mirrored backs and pull-out drawers. Two full showers and a restroom provide necessary amenities nearby for the performers.

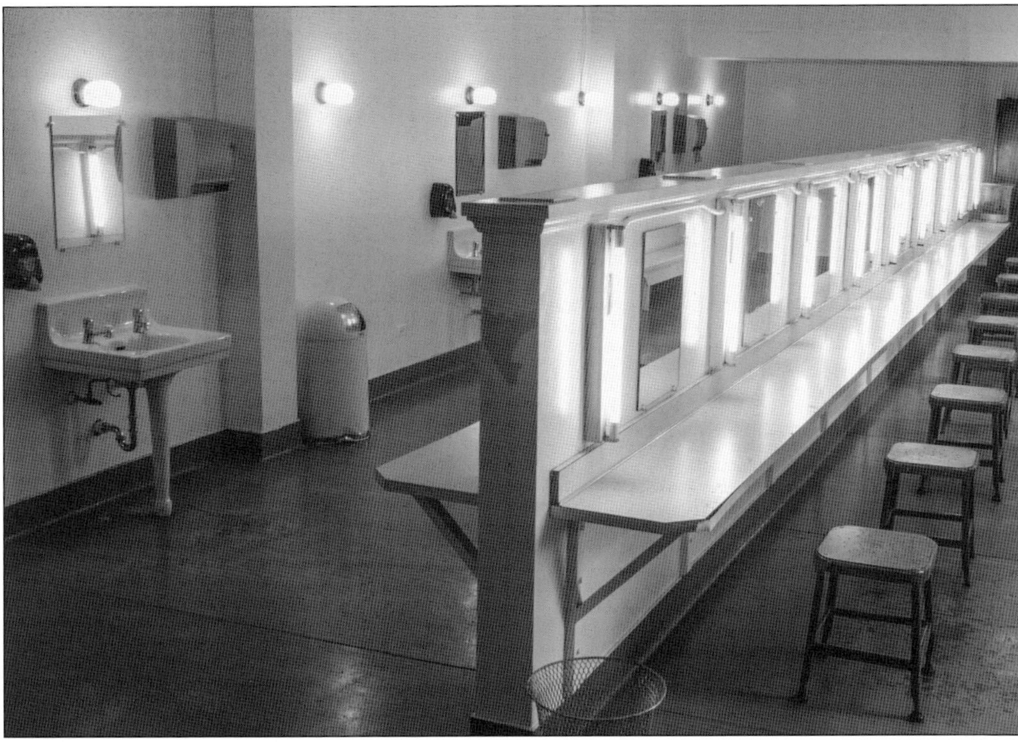

Makeup Stations. Twenty makeup stations and three sinks allow the actors and performers an area to ready themselves for the shows. Each station provides individual lighting and a mirror and plenty of room in which to maneuver.

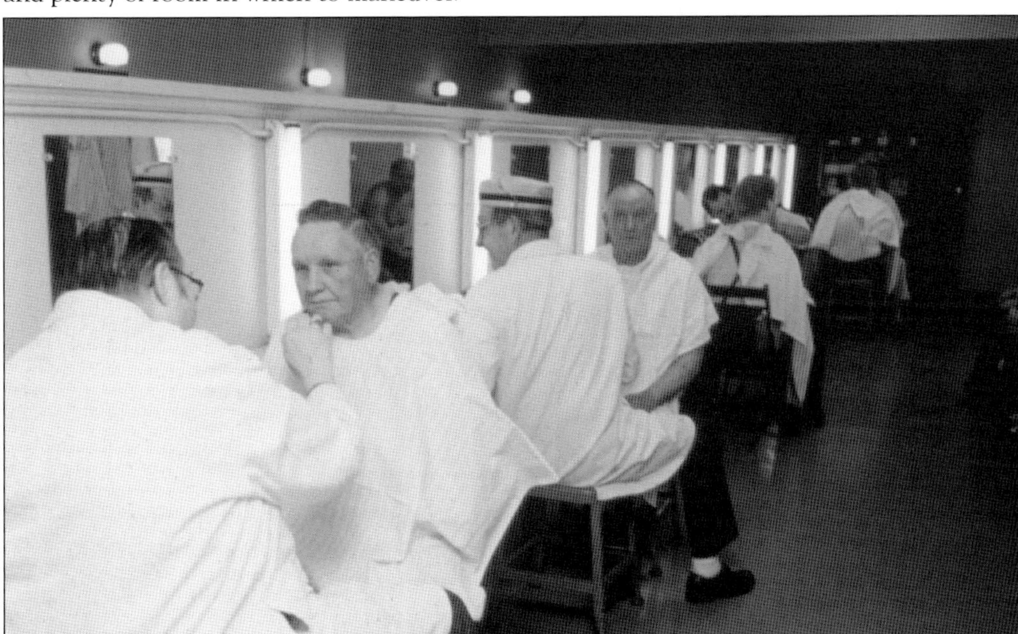

Makeup Artists. Worker hats stamped "Makeup" identify four Masonic members applying beards, wigs, and grease makeup to add color and realism to the actors about to perform.

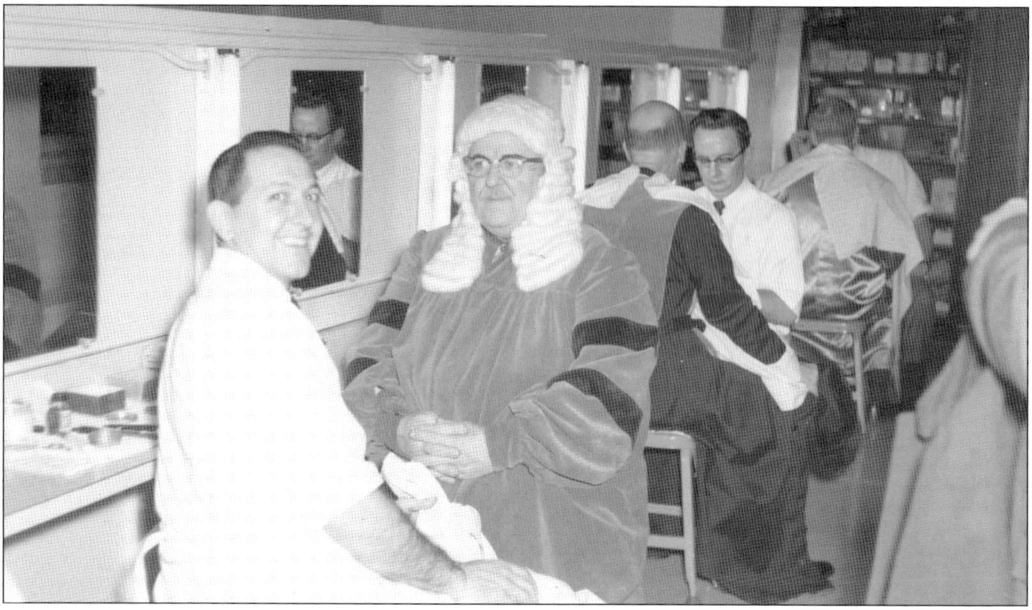

MAKEUP AREA. A judge, after leaving wardrobe, sits at a station in the makeup room while an artist plans out the finishing touches to make the actor look as realistic as possible while on the stage.

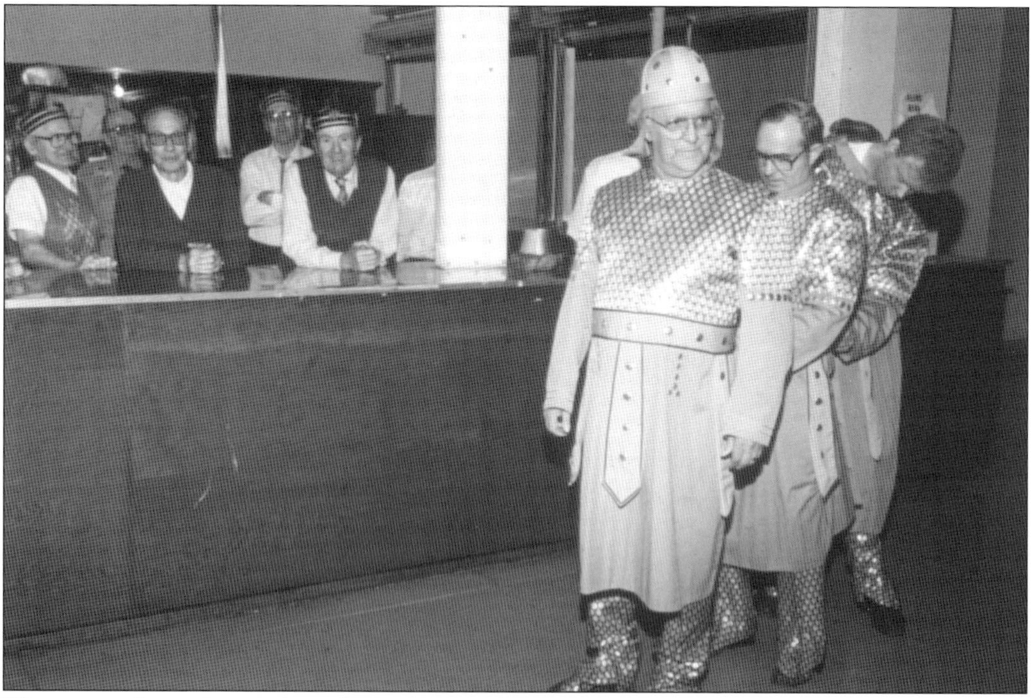

WARDROBE AREA. Located on the opposite side of the building from makeup, wardrobe houses hundreds of costumes and wigs valued at $82,000 by the Industrial Appraisal Company. Pictured standing from left to right, three cast members (Sidney Goline, Scott Heilman, and Jeff Jackson) adjust each other's 32nd-degree-guard outfits while wardrobe experts (including Dick Carns) stand in the background admiring their work.

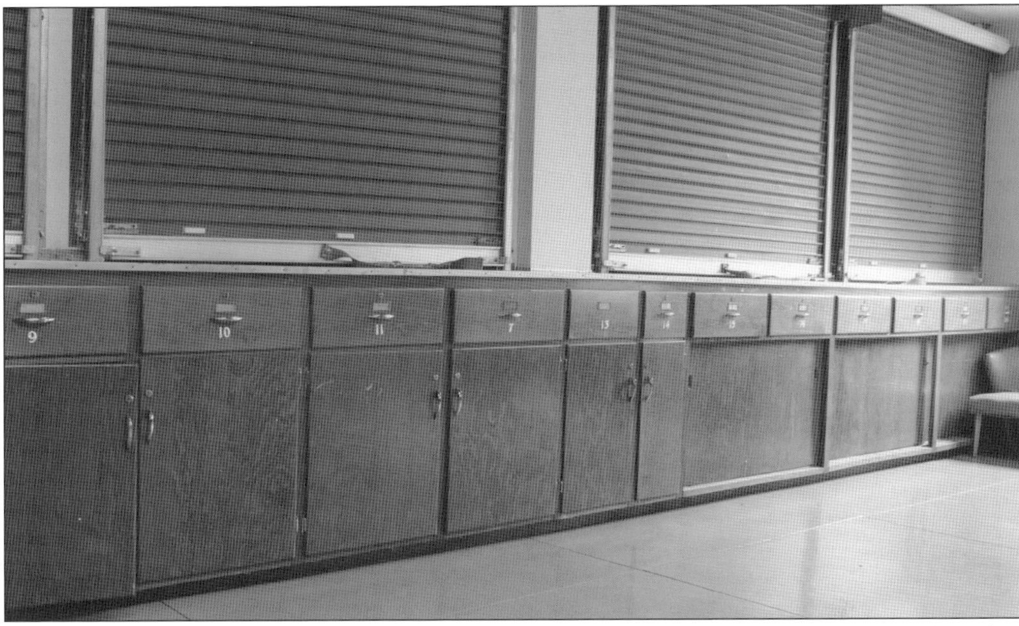

Wardrobe Room. The wardroom room's 20 drawers house Masonic aprons, gloves, hats, lace bibs, colonial shoes, extra buckles, and white leggings. The fraternity stores an array of larger items such as military helmets, colonial flags, a supply of beards, and sandals in cabinets directly below the drawers.

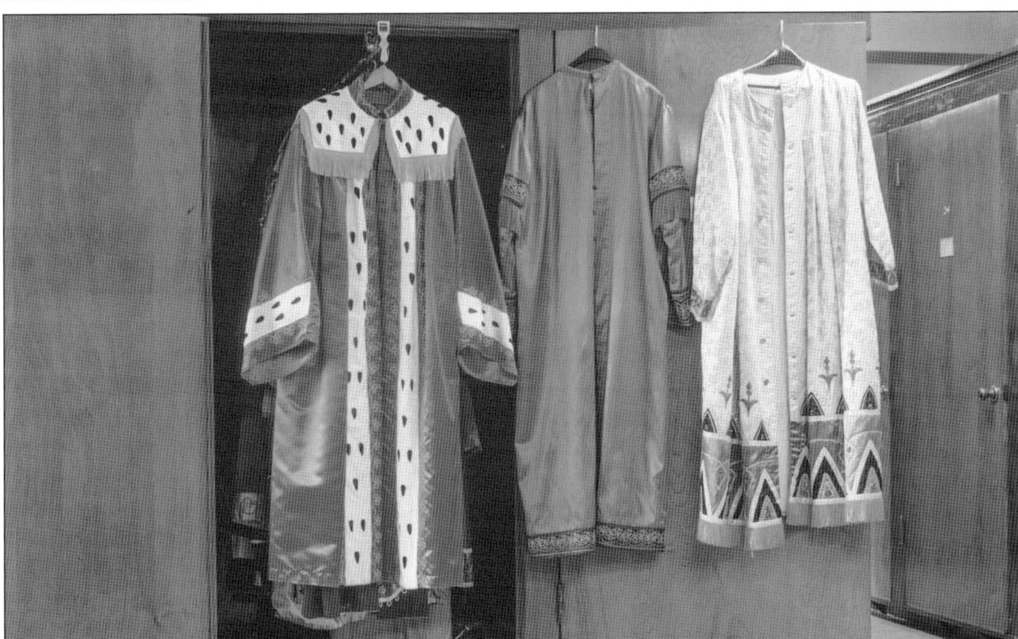

Old Testament–Era Costumes. Twenty-eight closets in the wardrobe room house hundreds of costumes for degree work. These outfits replicate what would have been worn around 530 BC during the rebuilding of the temple in Jerusalem. Key characters in the degree work include Israelite leader Joshua; Governor Zerubbabel of the province, later known as Judah; and King Nebuchadnezzar of Persia.

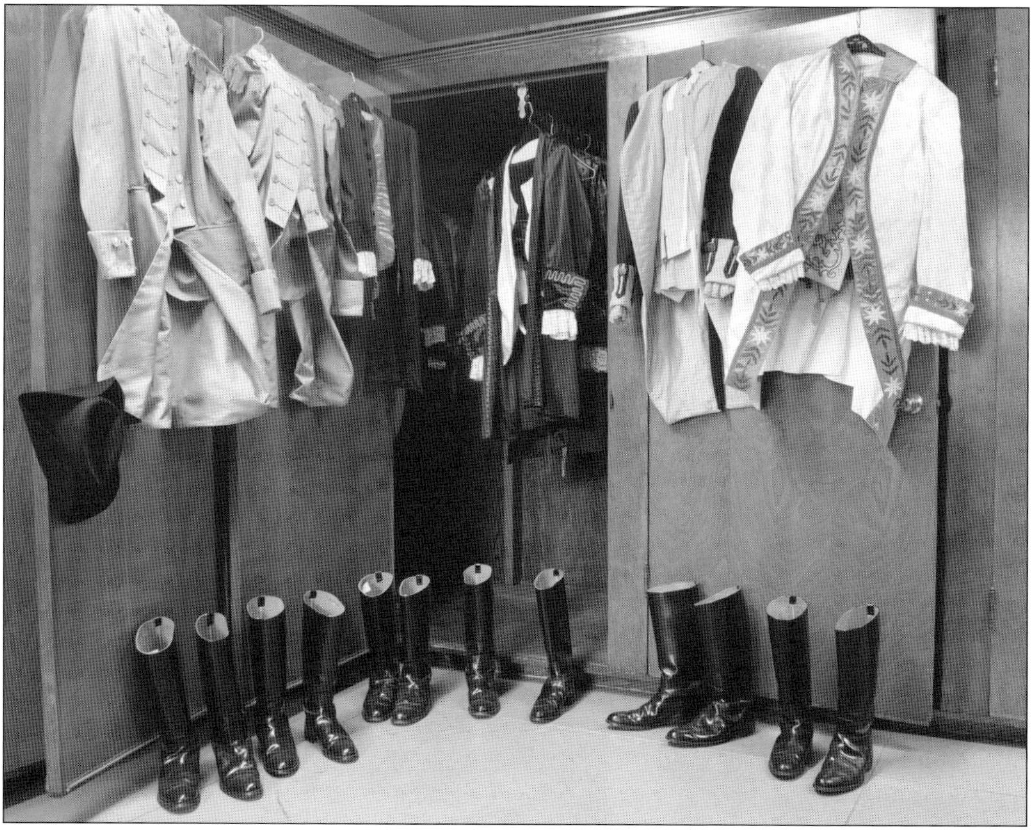

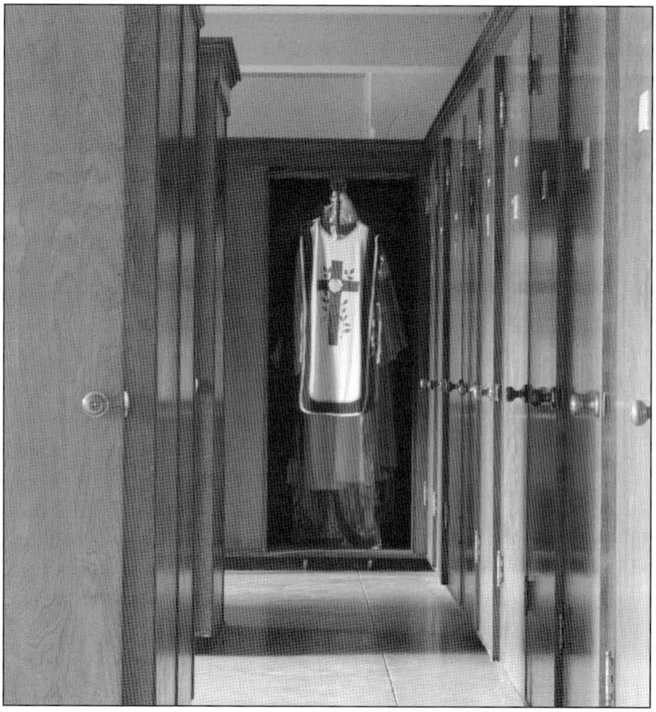

Colonial-Era Costumes. Three cabinets hold some of the high-topped boots, tricorne hats, and ornate costumes of the colonial era worn by Gen. George Washington and a number of Revolutionary War heroes. Powdered wigs, lace neck bibs, metal shoe buckles from the era, and Masonic regalia taken from the drawers add the necessary detail to bring these costumes to life.

Rose Croix Robe. This knight's garment hangs ready for use in the 18th degree, which affirms the broad principles of universality and tolerance. Each of the cabinets holds costuming for various degrees.

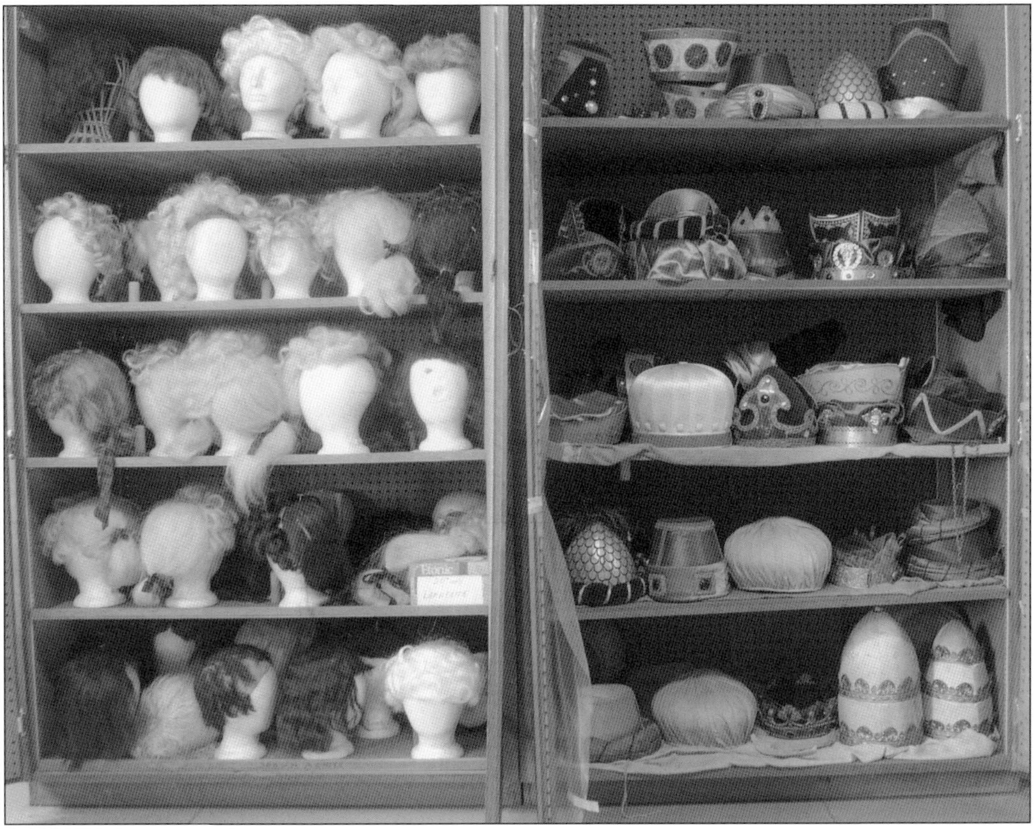

Wigs and Headwear. Cabinet numbers 29 and 30 house a variety of wigs primarily used for colonial degree work and a small portion of the hats, helmets, and jeweled crowns from the Old Testament era. Many of the pieces reflect elaborate workmanship, making them difficult and expensive to replace.

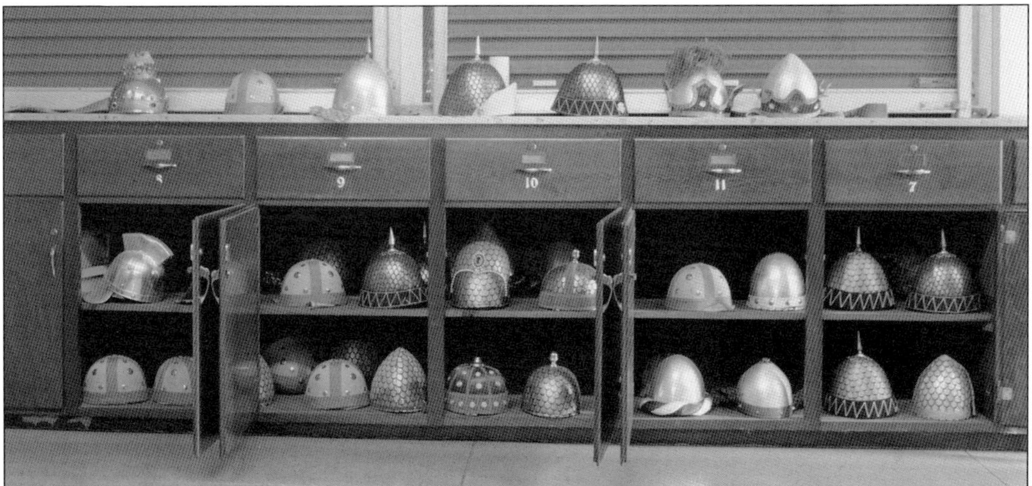

Helmets for Degree Work. Photographer Bill Conkle performs his magic to capture this case filled with dozens of military and regal headpieces used in a variety of degree work and special presentations such as *Darius*.

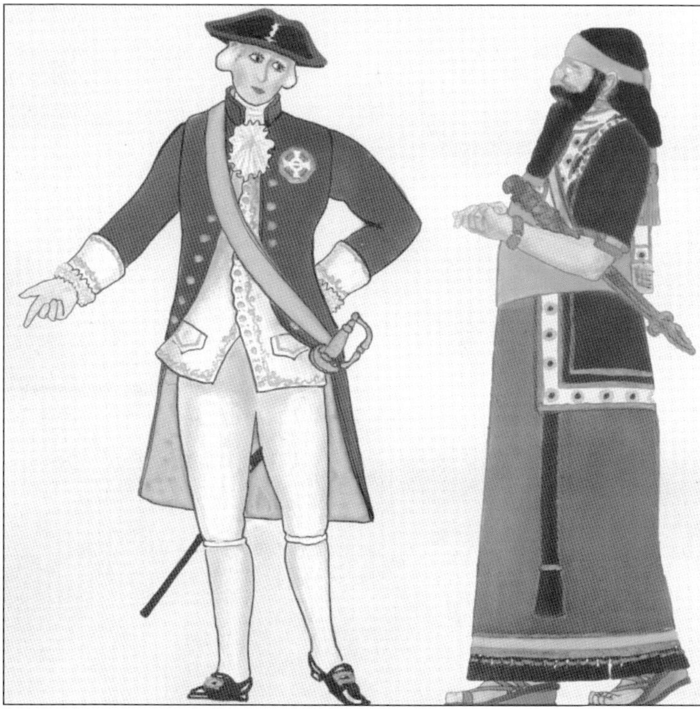

Artist Renditions Used by Costuming. The Masons take costuming seriously. Dozens of descriptive drawings present exemplars on which actual costumes are based. These pictures reflect two different periods utilized in degree work. Each detail is carefully described. For example, the reverse side of the colonial officer sketch reads, "A blue velvet coat with gold buttons, a white satin cuff with frilled ruffles at the wrist attached to the inside of the coat sleeve, highlighted by gold edging" as part of the description.

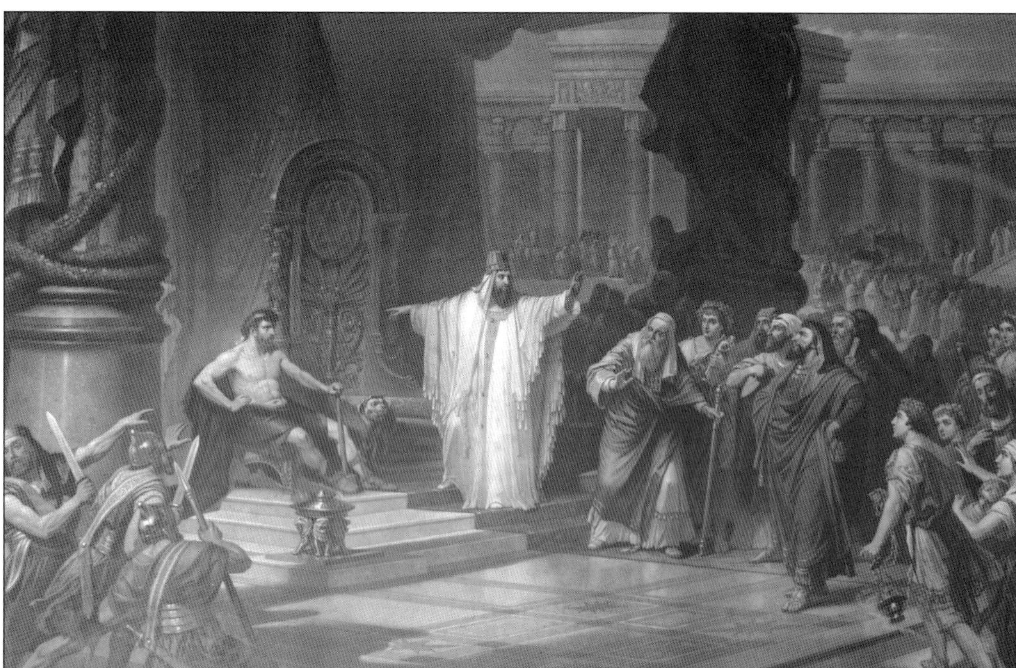

Ironworker and King Solomon. John Sartain created this illustration in 1871, found in the Wallace Library. Pictures like this served as guidelines for costuming used in Masonic degrees.

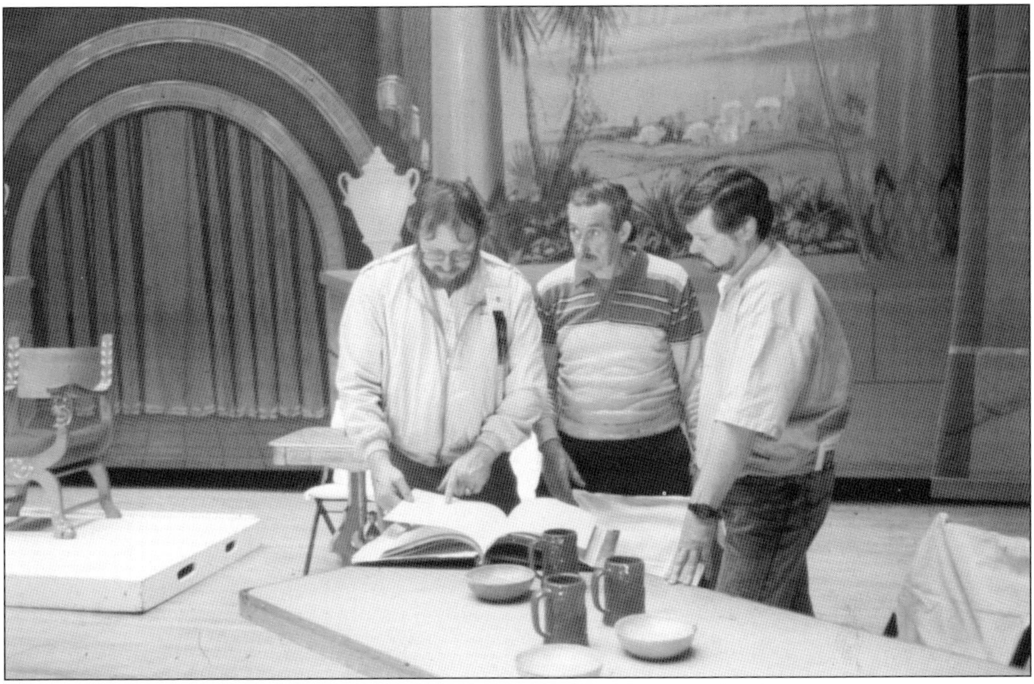

Degree Preparation. Stage crew members collect and line up props for the upcoming presentation while the actors rehearse their lines in another area. The team lays out each item in strict accordance with the notations found in the instruction book.

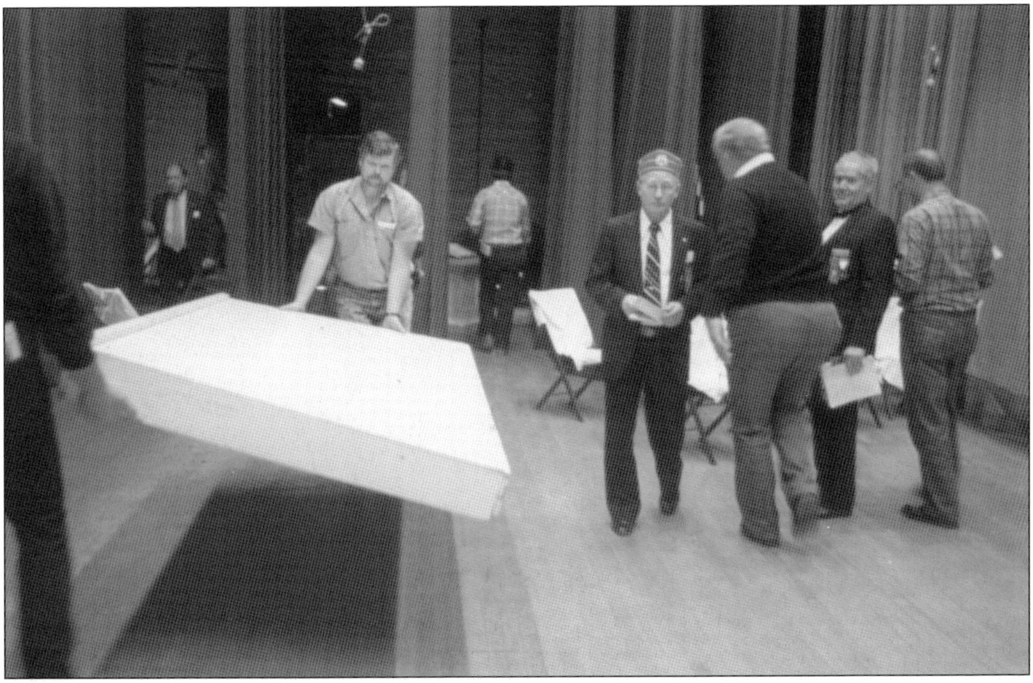

Setting the Scene. Stage manager Dave Smith oversees the positioning of the riser to its proper location. The crew must consider dozens of details to insure the accuracy and effectiveness of every scene.

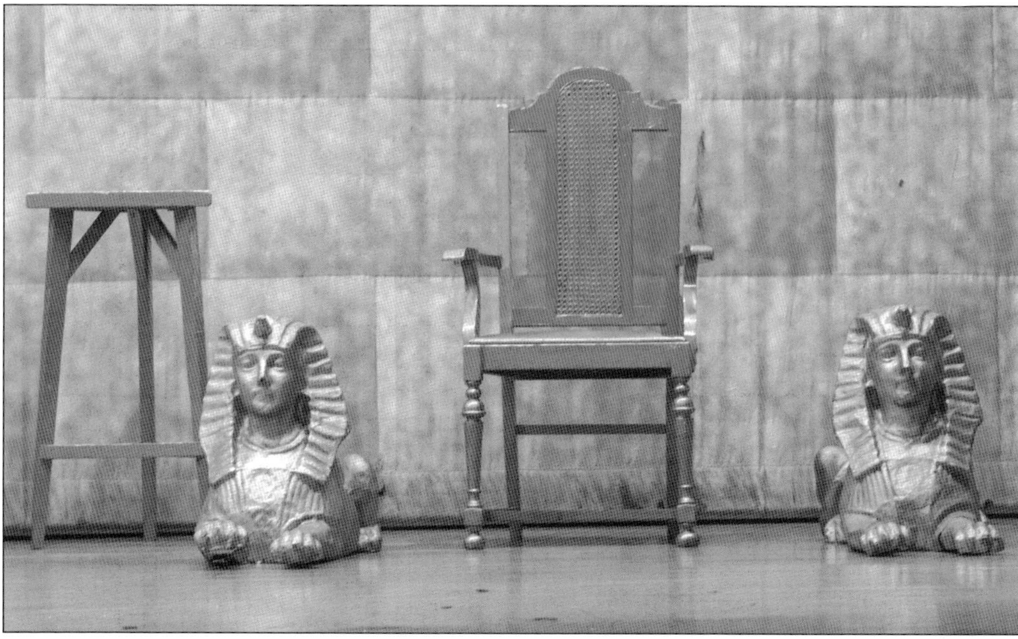

Golden Sphinx Prop. Some of the hundreds of unusual props used for degree work include this pair of painted, papier-mâché sculptures, each one with the head of a woman planted on the body of a lion.

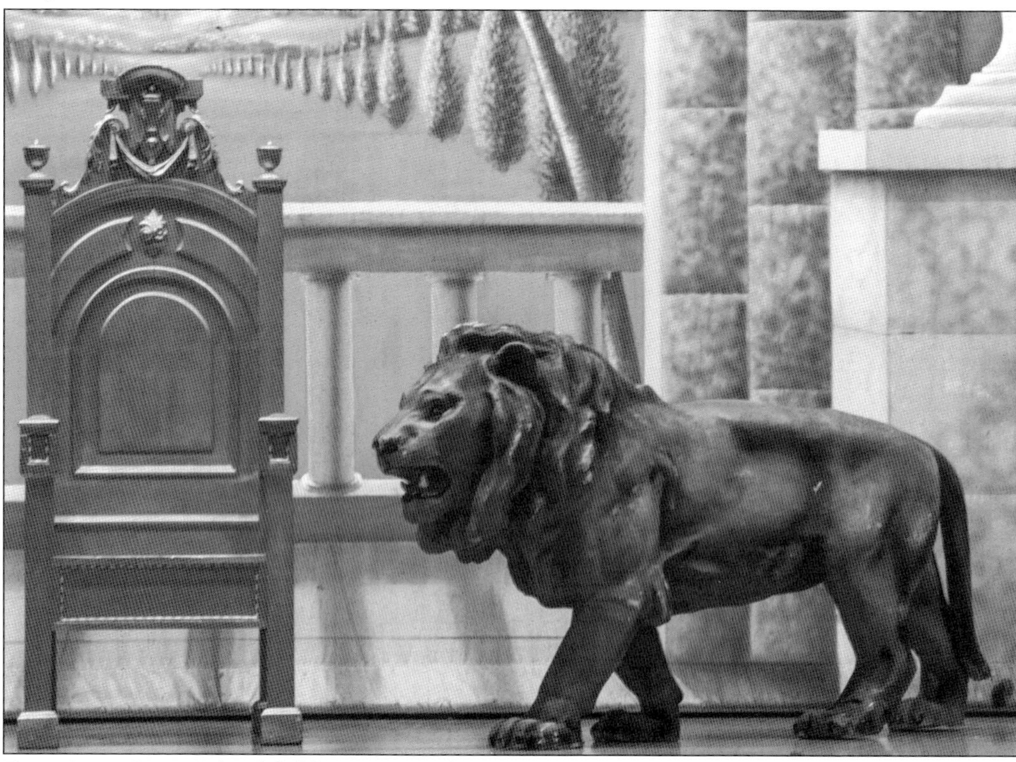

Full-Sized Lion. This life-like, tawny-colored, snarling-male lion adds realism to any set. The glass eyes, white teeth, and curly mane present an imposing sight.

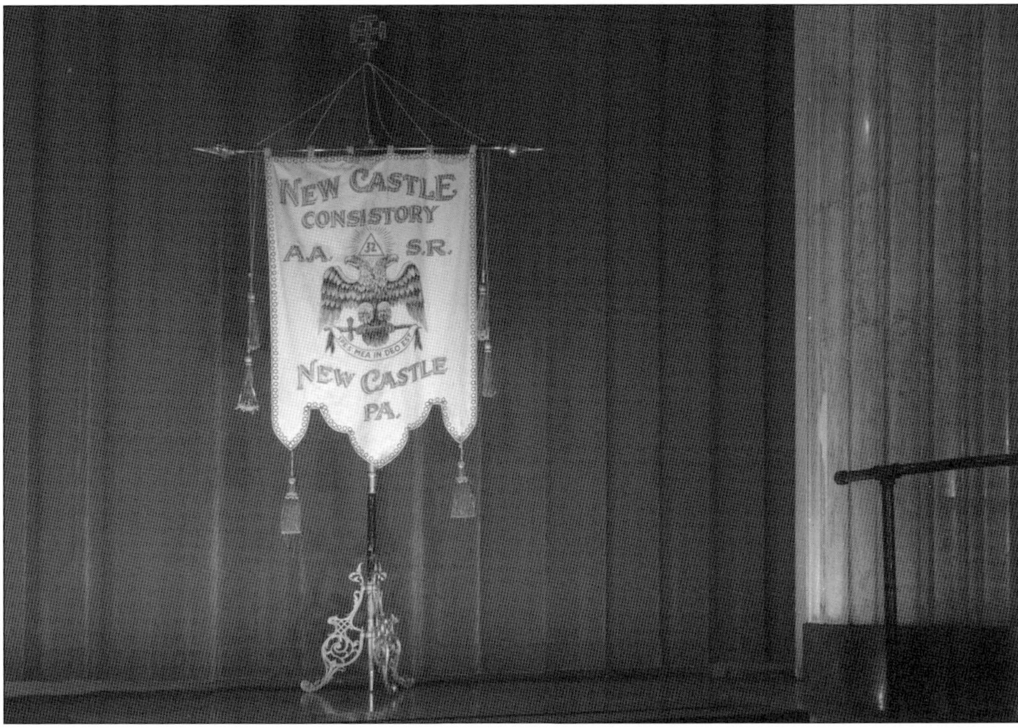

NEW CASTLE CONSISTORY BANNER. A gold and black tripod flagstaff holds a brocaded, screen-printed banner with a double-headed eagle whose talons hold a sword draped with the Consistory Latin motto reading, "Spes Mea in Deo Est," translating to "My Trust is in God." The initials "A.A.S.R." above the eagle stand for "Ancient and Accepted Scottish Rite." A Maltese Cross and multiple tassels tie the display together and add richness.

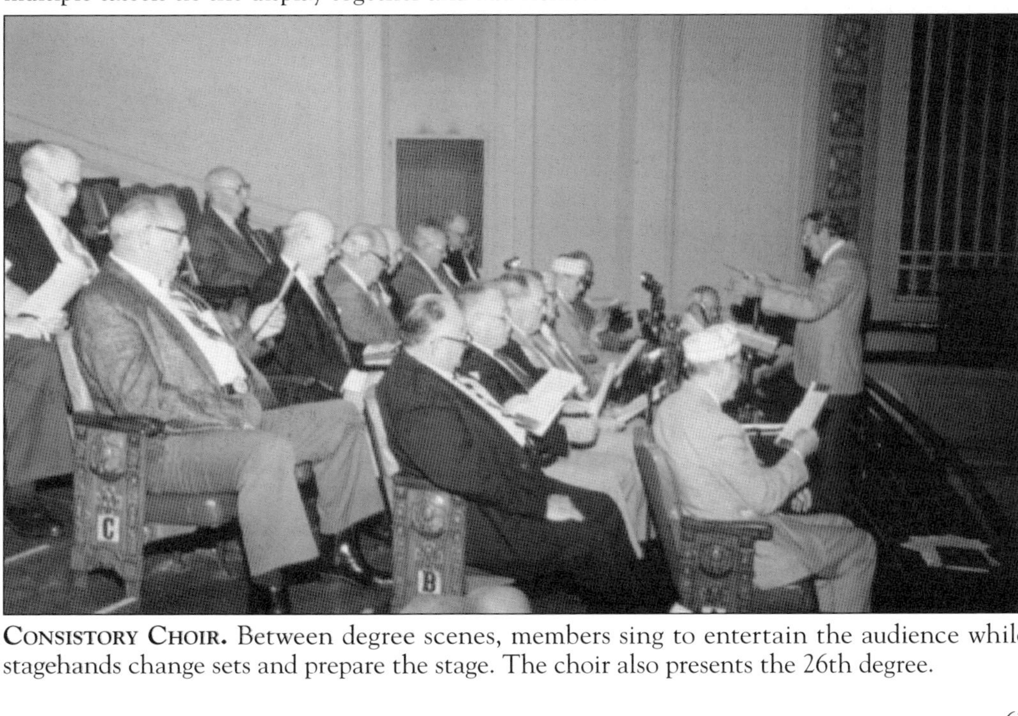

CONSISTORY CHOIR. Between degree scenes, members sing to entertain the audience while stagehands change sets and prepare the stage. The choir also presents the 26th degree.

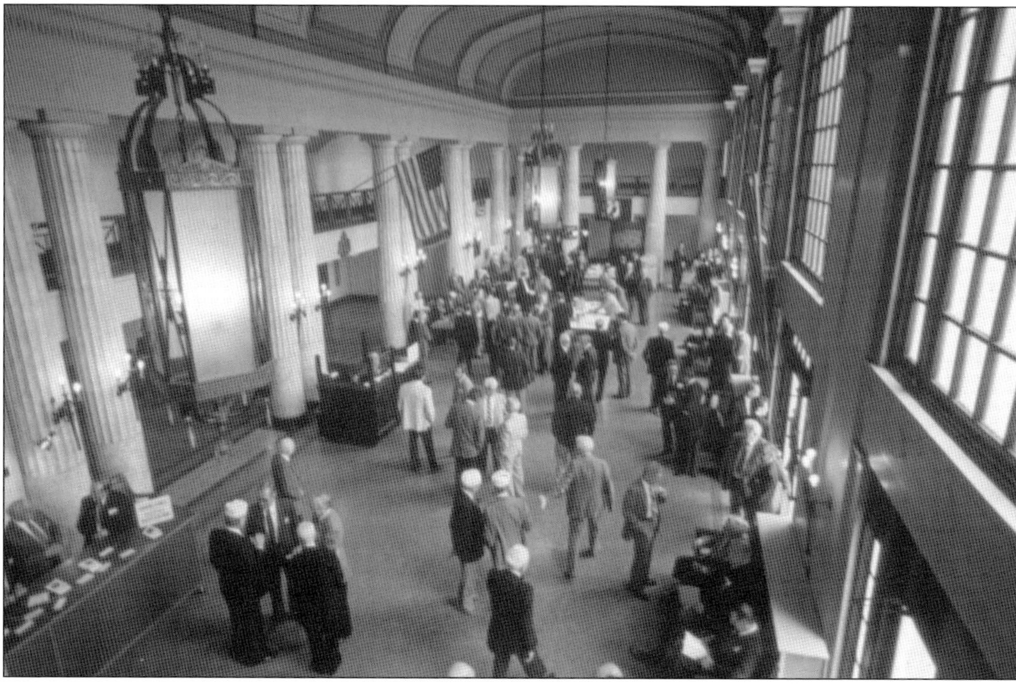

REUNION ATTENDEES. Candidates and members fill the lobby to register for the day's reunion activities. The Legion of Honor and other Masonic groups explain their functions to interested parties.

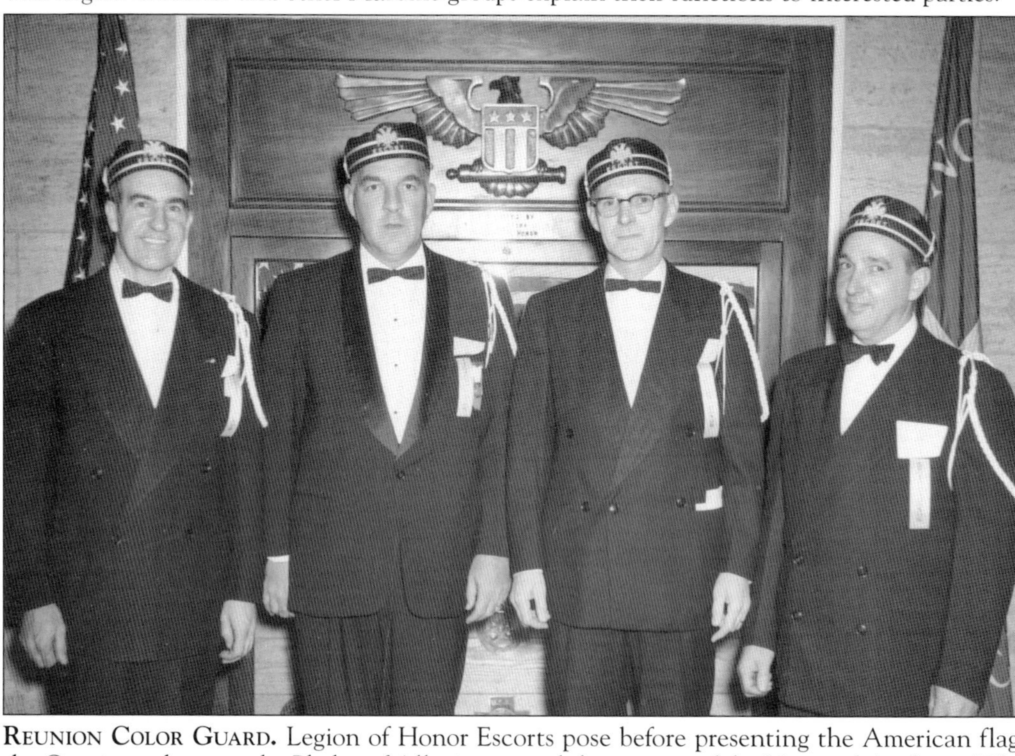

REUNION COLOR GUARD. Legion of Honor Escorts pose before presenting the American flag, the Consistory banner, the Pledge of Allegiance, and the singing of the National Anthem, "The Star-Spangled Banner."

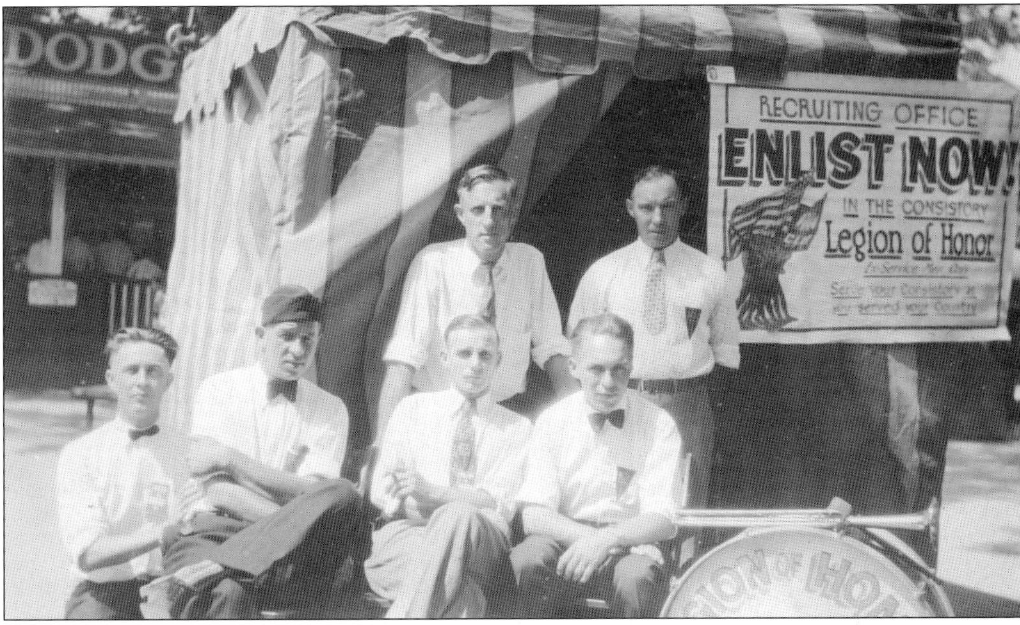

Legion of Honor. Members of the Legion of Honor, all of whom had served their country in the US Armed Forces, enlist new recruits during their annual Masonic outing at Cascade Park. The dodgem thrill ride can be seen in the background.

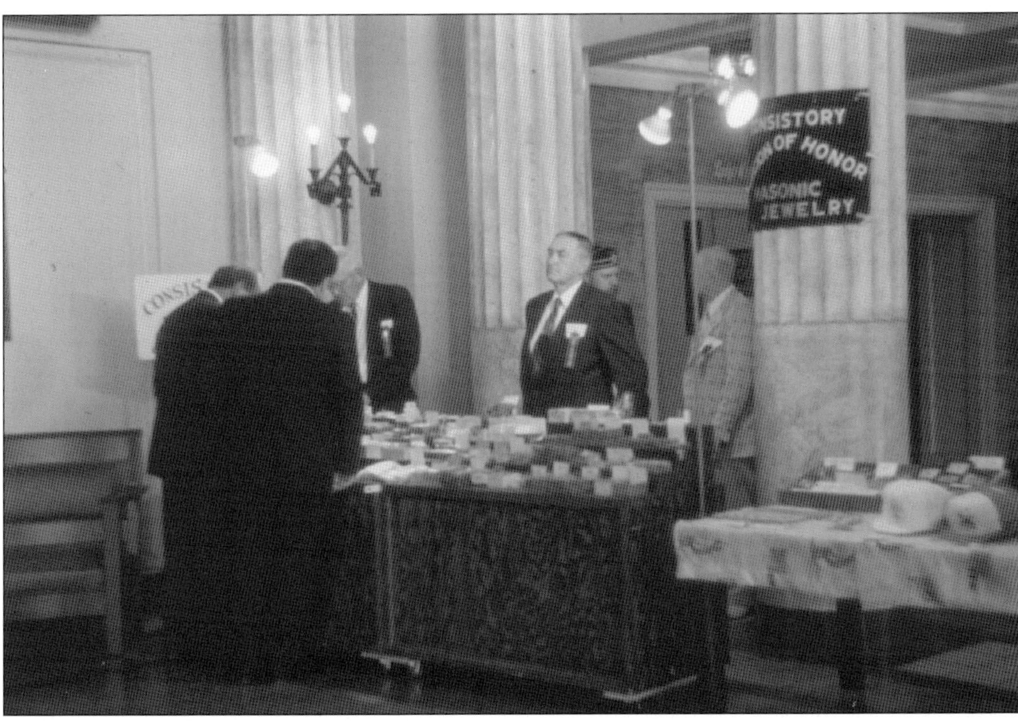

Legion of Honor Merchandising. Volunteers standing beside the lobby pillars do a brisk business selling jewelry, hats, and Masonic paraphernalia to eager members at a reunion.

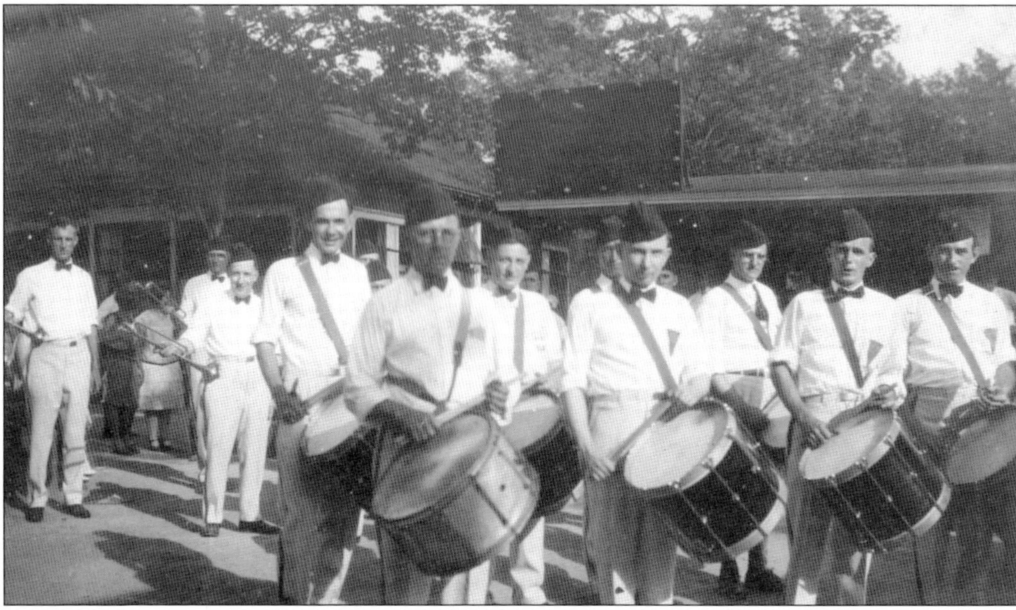

Drum and Bugle Corps. Musicians from the Consistory Legion of Honor line up for a parade at Cascade Park. Members received free ride passes and meal tickets for themselves and their family.

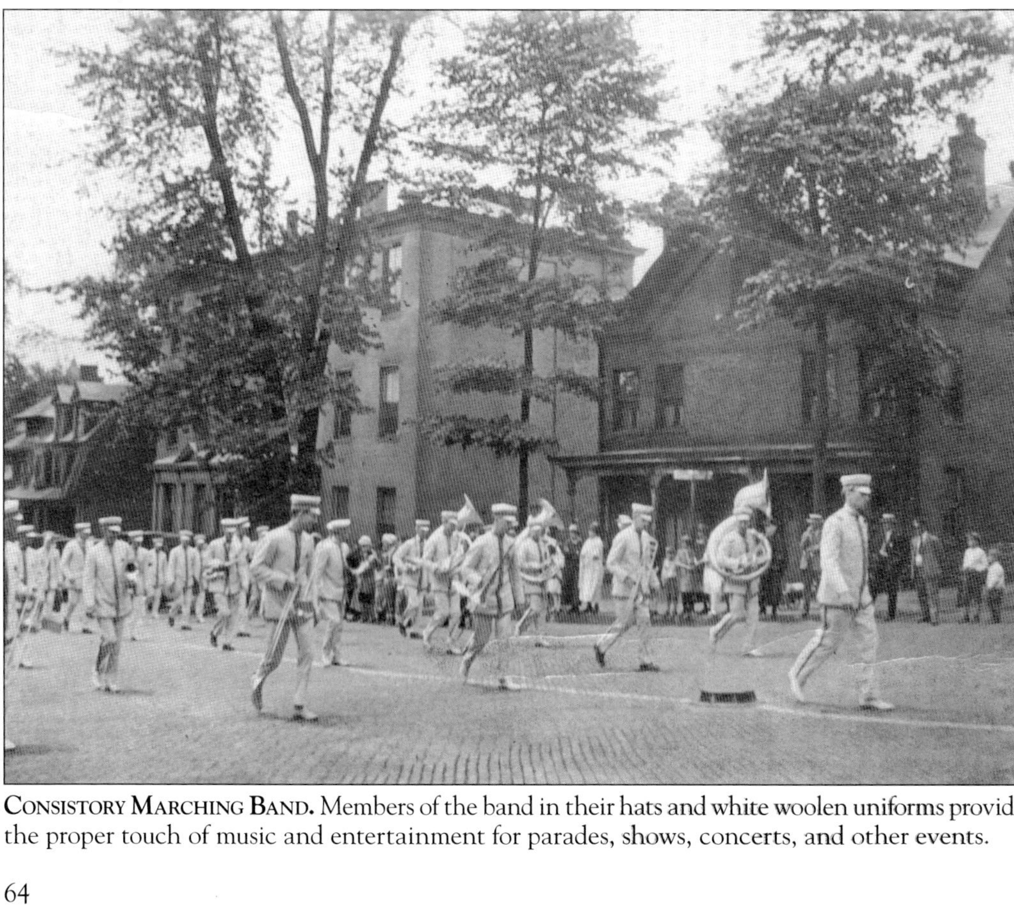

Consistory Marching Band. Members of the band in their hats and white woolen uniforms provide the proper touch of music and entertainment for parades, shows, concerts, and other events.

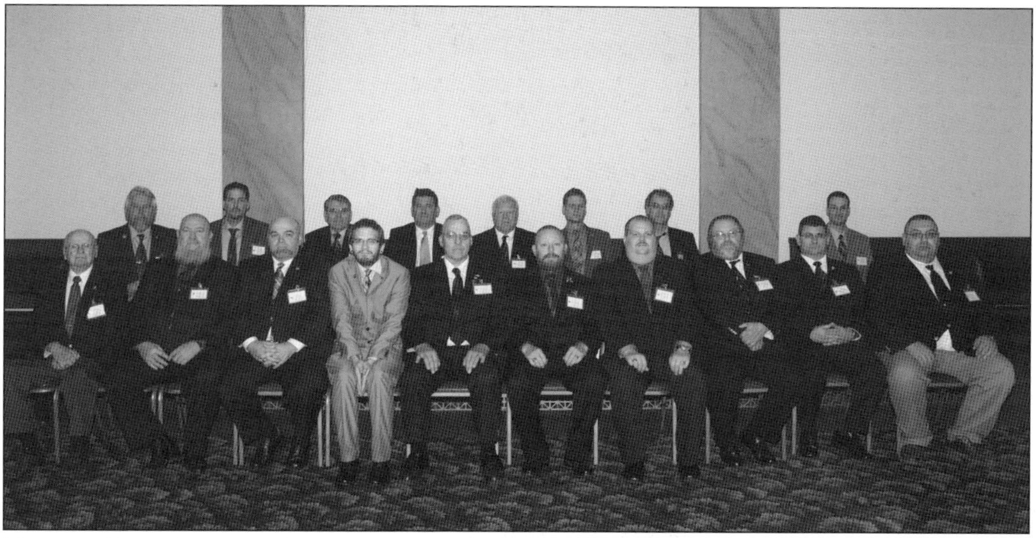

The 2015 Fall Reunion Class. New members join in a one-day class adding to the ranks of the Valley of New Castle Consistory, the 214th such reunion for the organization.

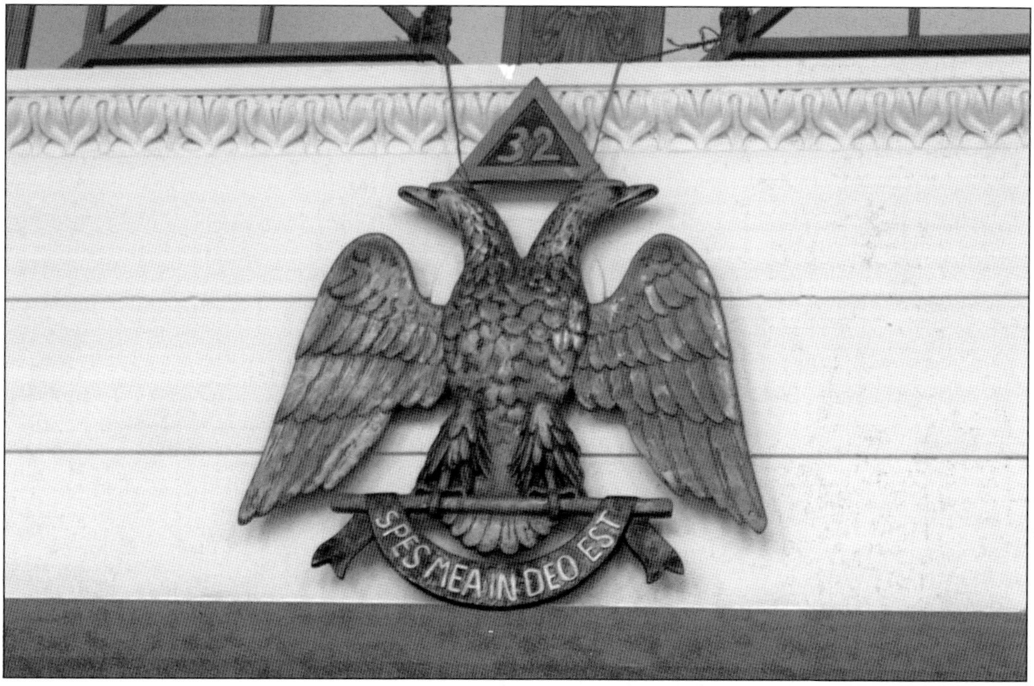

Double-Headed Eagle of Lagash. This symbol, which hangs from the mezzanine level of the lobby, stands as the oldest royal crest in the world, going back more than 2,000 years before the building of King Solomon's Temple. Freemasonry first used the eagle in Paris in 1758, and it has remained an important part of the Consistory ever since.

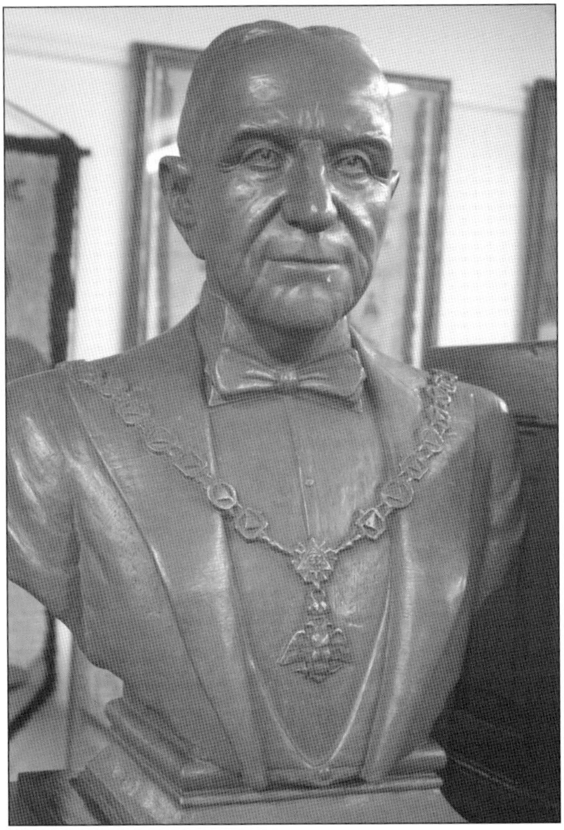

BUST OF JOHN S. WALLACE. The Supreme Council constituted the Valley of New Castle on October 15, 1909, with John Scofield Wallace, 33rd degree, as the first Thrice Potent Master. The library also contains a large oil painting of Wallace valued by the Industrial Appraisal Company of Pittsburgh at more than $4,000.

LADIES' AFTERNOON TEA. Candidates' wives enjoy tea, hand-made cookies, and a pleasant get together in the East Lodge Room while their husbands receive degree work in the auditorium. Generally, the women toured some interesting local spot such as the Shenango Pottery prior to the tea.

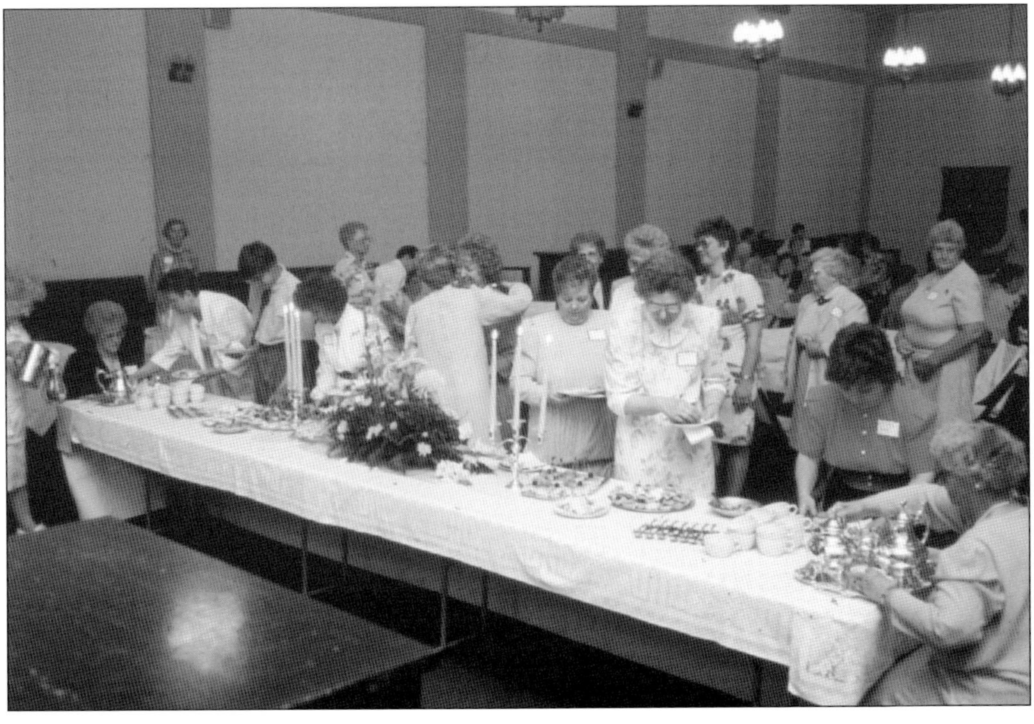

Four

Appendant Organizations

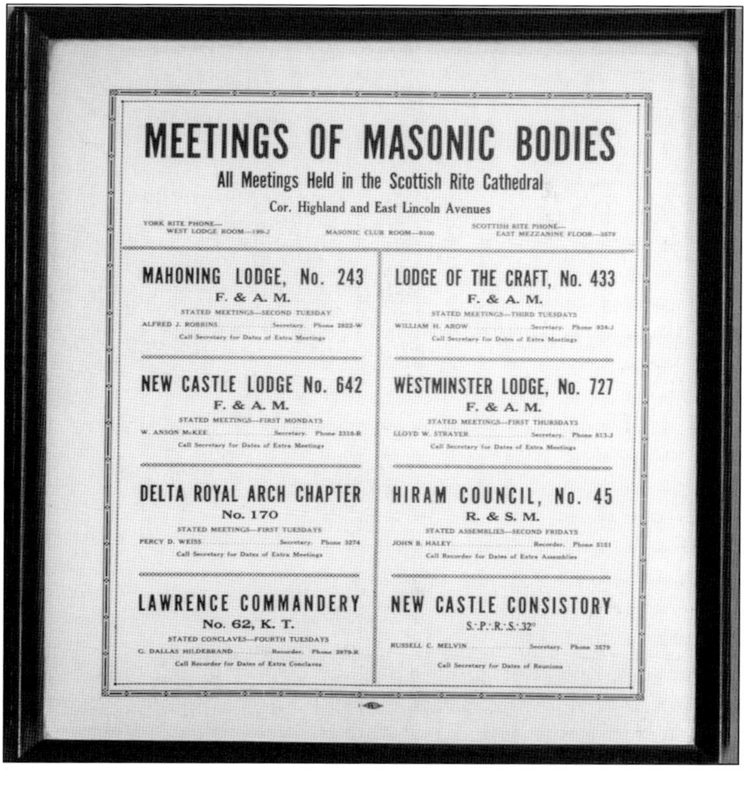

Scheduled Meeting Times Sign. New Castle had four Blue Lodges that met monthly: Mahoning, New Castle, Westminster, and Lodge of the Craft. Three York Rite bodies and one Consistory chapter also met on a regular basis.

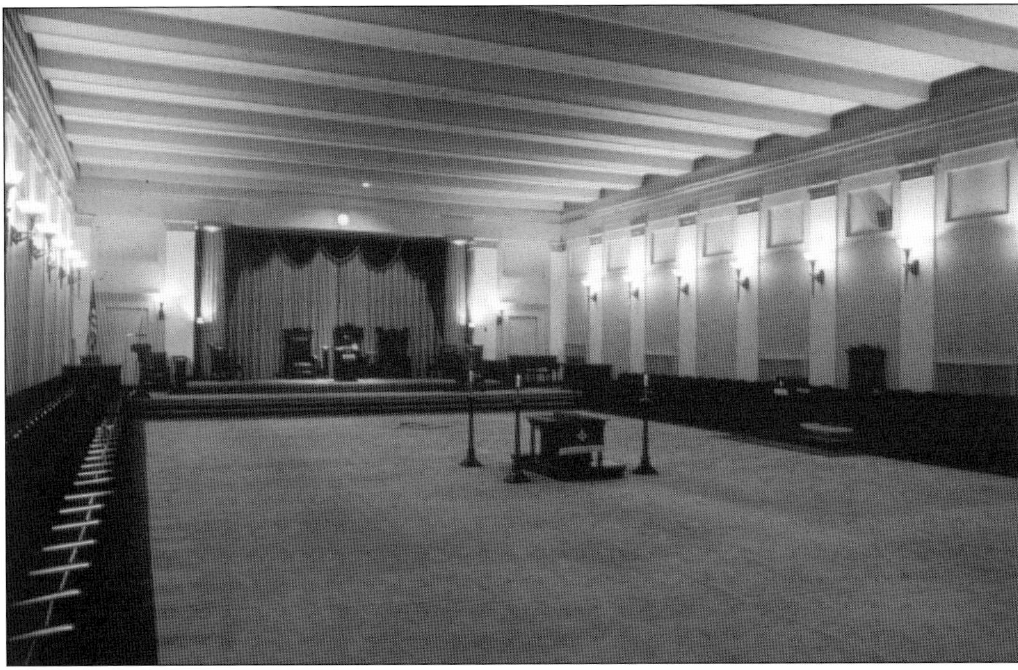

West Lodge Room. The brothers have set up the area for degree work or a lodge meeting. The center altar contains the Bible and a square and compass, symbolic emblems of Freemasonry. The station of the Worshipful Master who conducts the Blue Lodge meeting lies to the east.

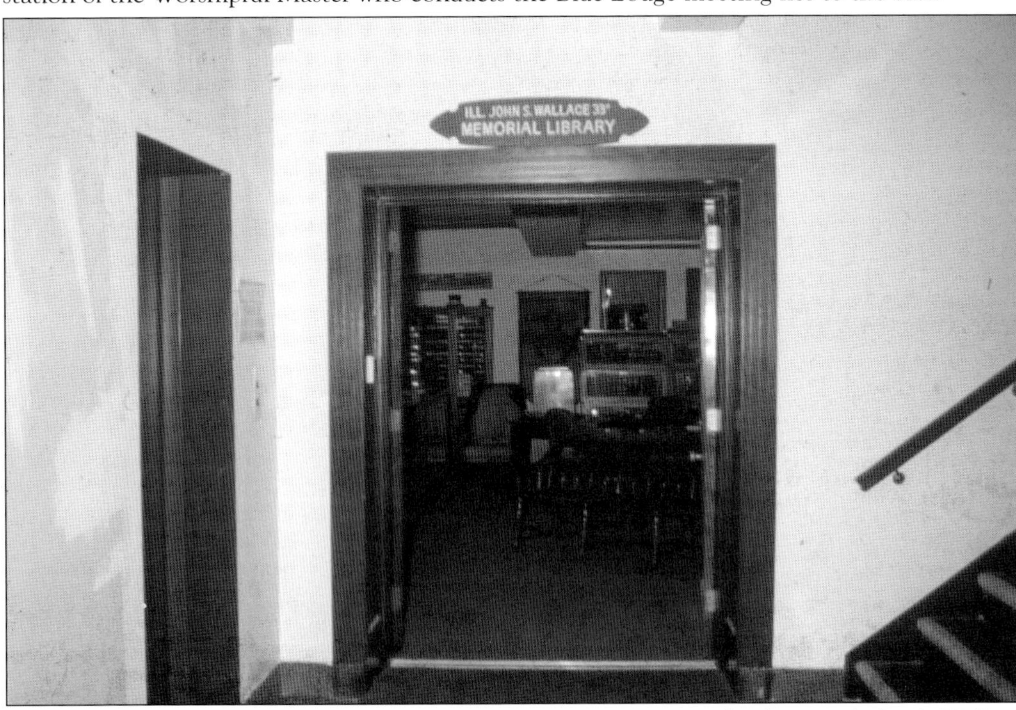

Wallace Library. This doorway leads to a library and museum named in honor of John S. Wallace, the first commander in chief of Valley of New Castle Consistory. The room is jam-packed with relics, antiquities, Masonic symbols, books, and other historical items.

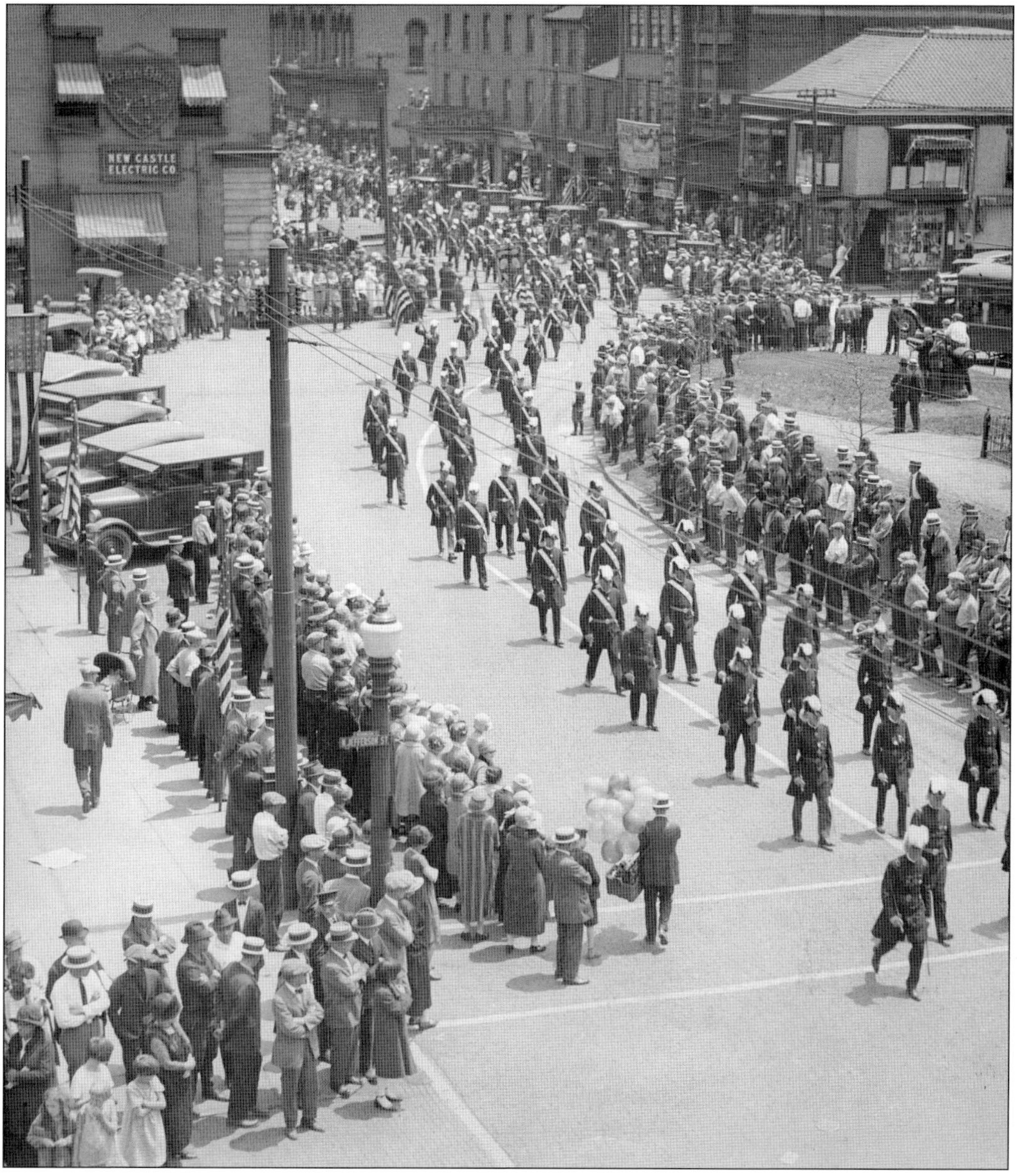

KNIGHTS TEMPLAR. Resplendent in their magnificent plumed hats and military uniforms, the Knights Templar members march through the city of New Castle during a parade.

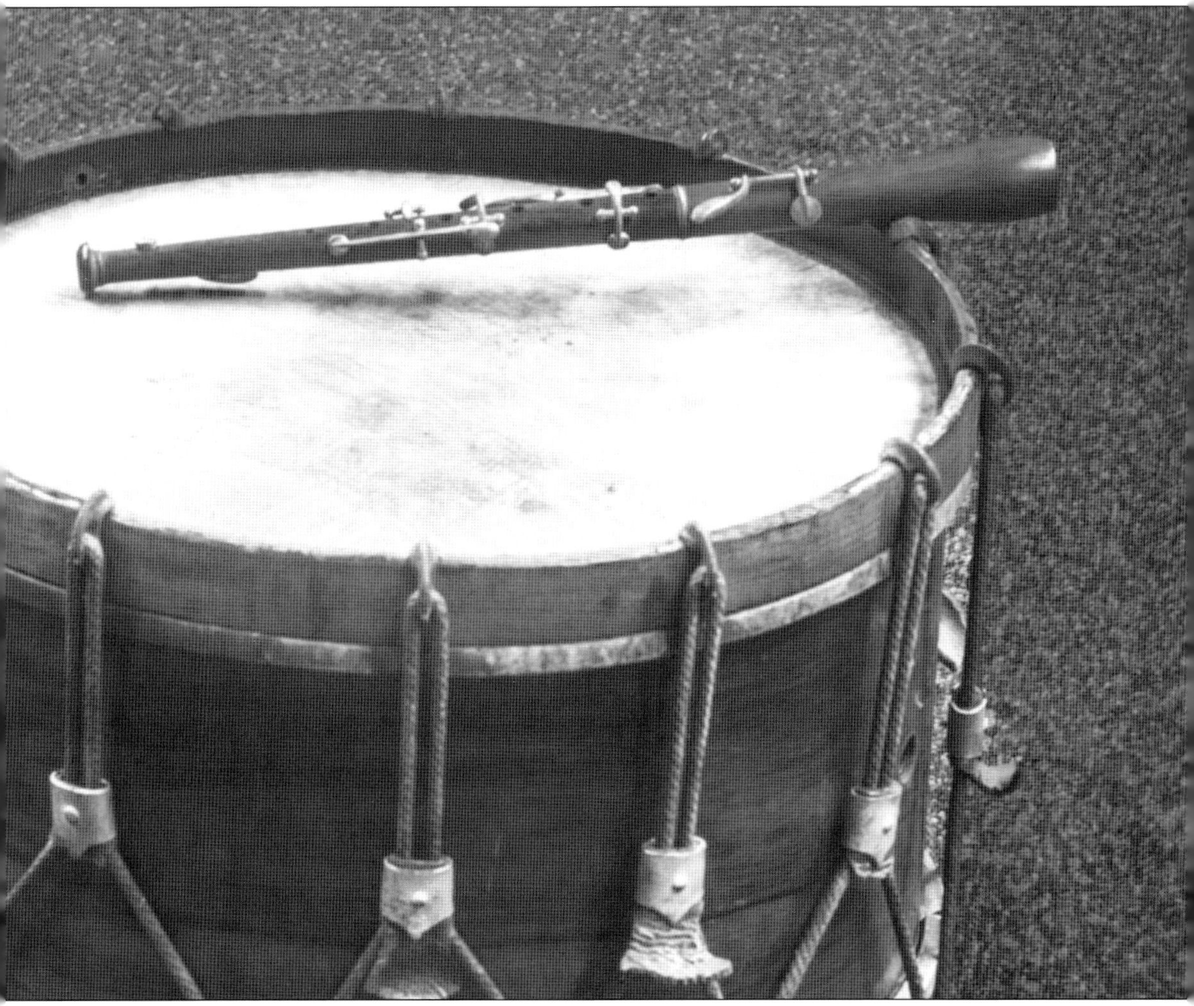

Antique Musical Instruments. This woodwind, which is not in playable condition, belonged to Dillon Crist, 33rd degree and Grand Treasurer of Ohio. Merle Leslie donated it to the library. The photograph also shows a snare drum in usable condition, once played by a member of the Consistory band. These musical instruments reflect a small sampling of the unique and unusual items to be found in the Wallace Library.

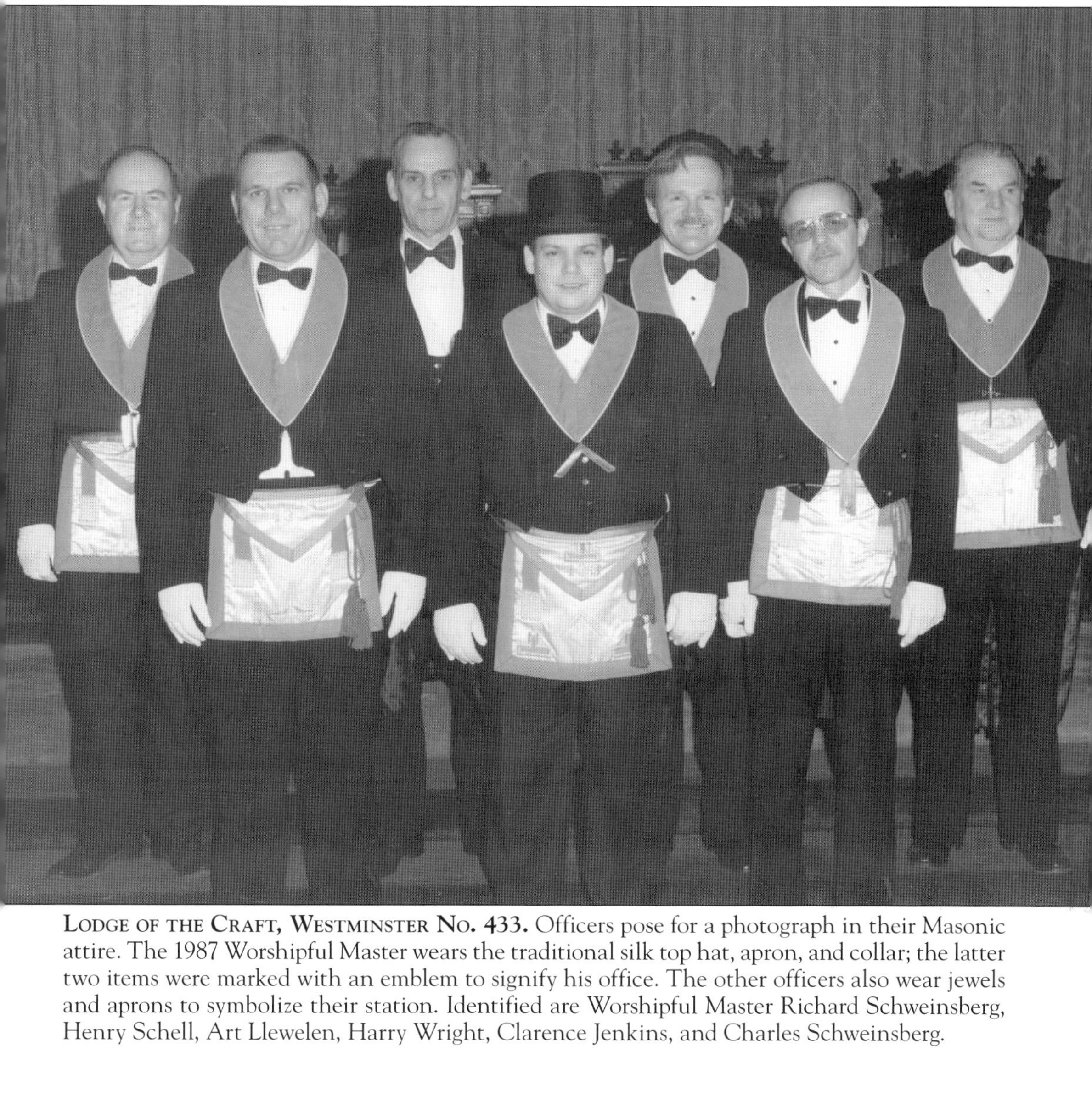

Lodge of the Craft, Westminster No. 433. Officers pose for a photograph in their Masonic attire. The 1987 Worshipful Master wears the traditional silk top hat, apron, and collar; the latter two items were marked with an emblem to signify his office. The other officers also wear jewels and aprons to symbolize their station. Identified are Worshipful Master Richard Schweinsberg, Henry Schell, Art Llewelen, Harry Wright, Clarence Jenkins, and Charles Schweinsberg.

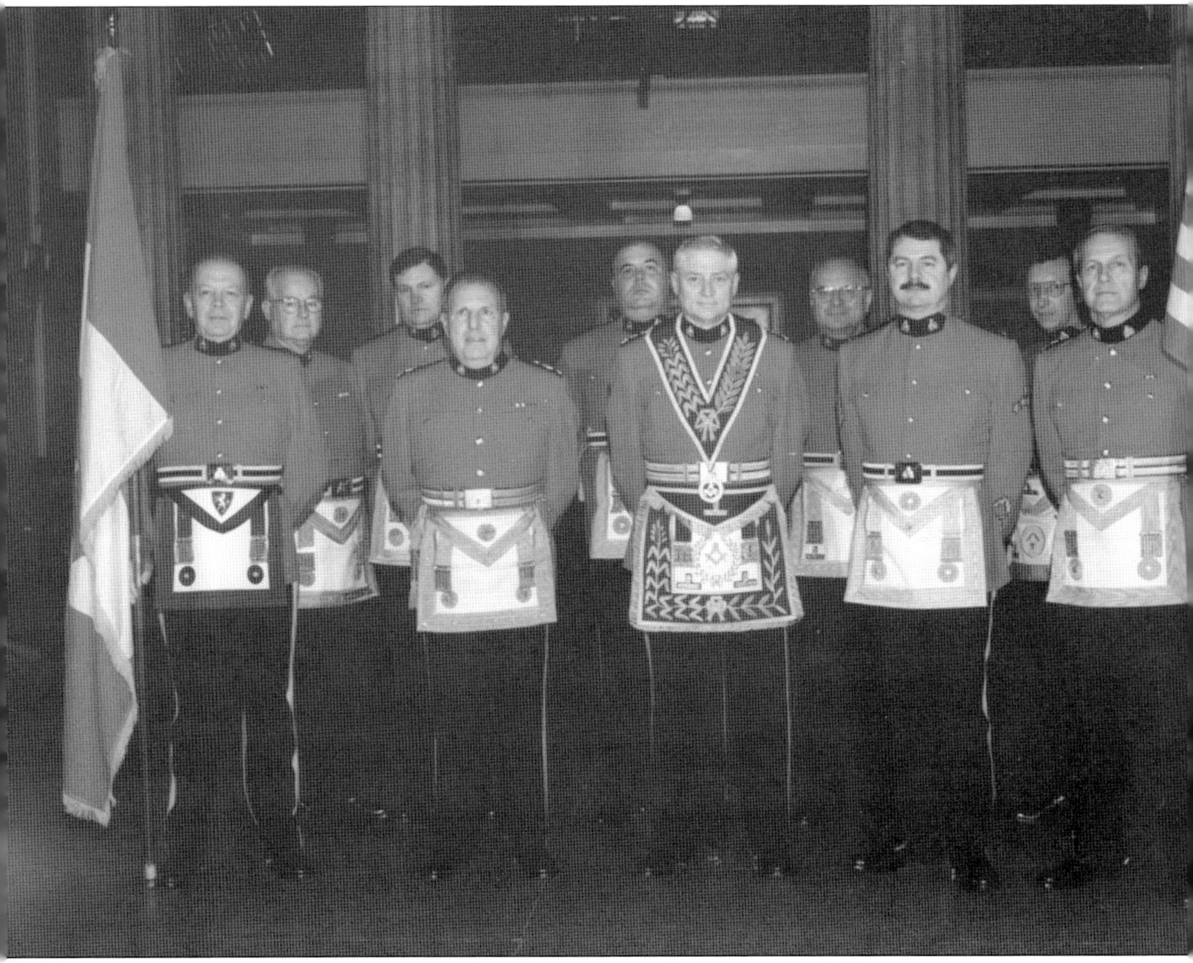

Masonic Guests. Canadian Royal Mounted Police visitors in the lobby prepare to join one of the New Castle Blue Lodges for a meeting. The officer in the center wears the jewel, collar, and apron representing a Grand Lodge. The Canadian flag joins the American flag in celebration. Freemasonry always welcomes their fraternal brothers from around the globe.

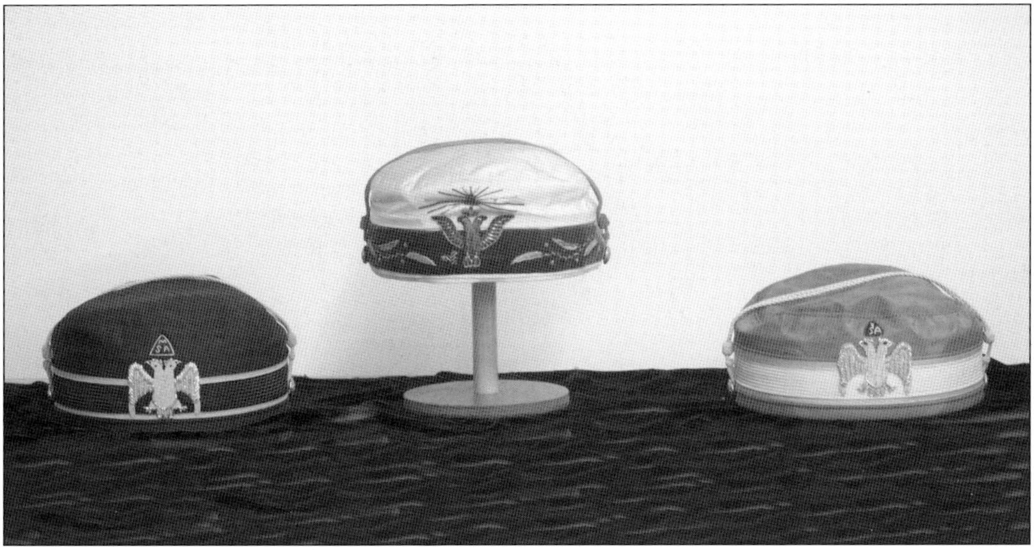

HONORARY SCOTTISH RITE HATS. A white hat denotes the wearer as a 33rd degree Mason. Meritorious Service Award recipients wear a red hat, and Valley Service Award selectees earn a green hat. The fraternity grants these prestigious recognitions on an honorary basis by invitation only.

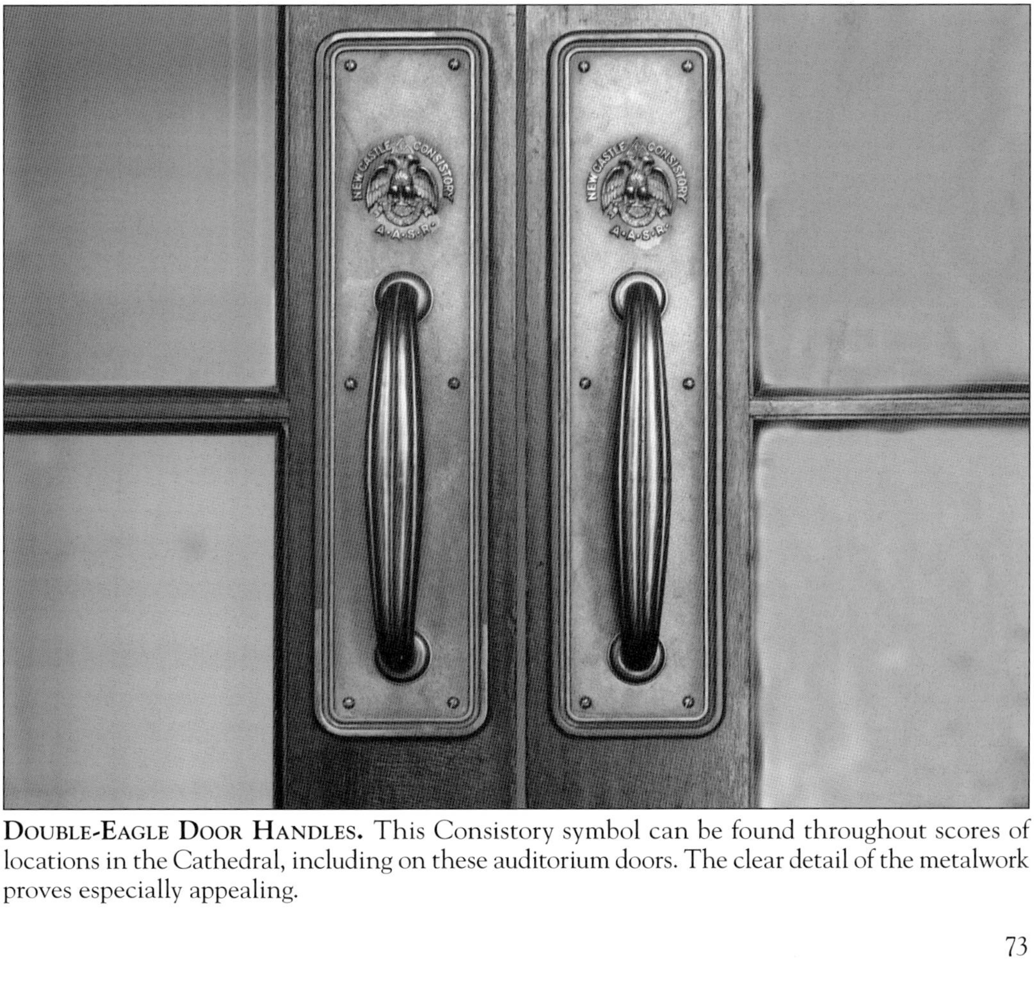

DOUBLE-EAGLE DOOR HANDLES. This Consistory symbol can be found throughout scores of locations in the Cathedral, including on these auditorium doors. The clear detail of the metalwork proves especially appealing.

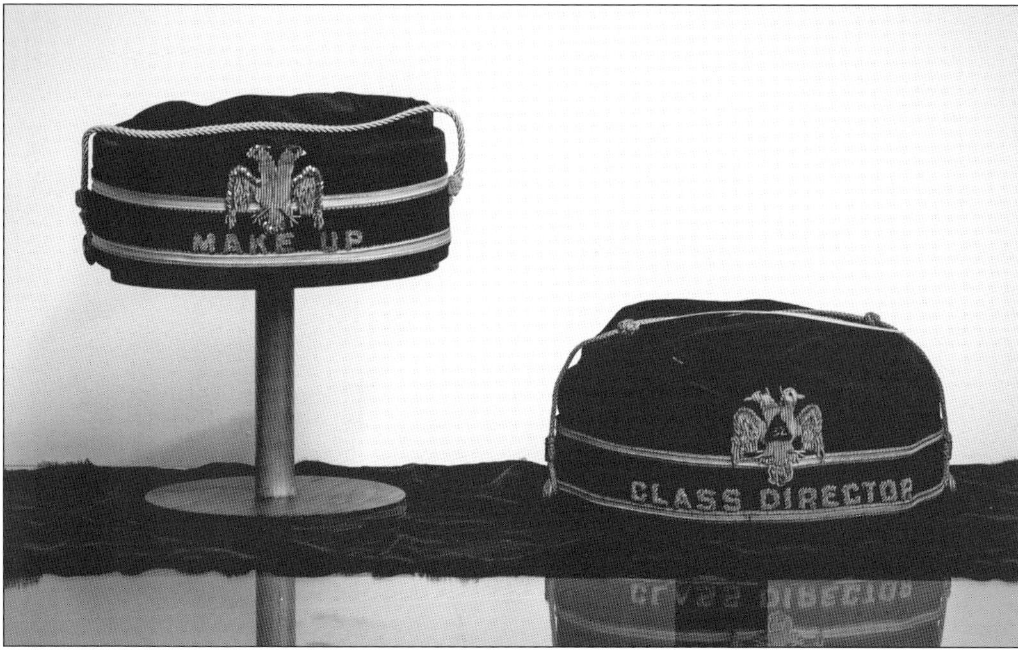

Valley Workers' Hats. Volunteer helpers wear these black hats during reunions to designate their roles. One can recognize a stage crew member, lighting operator, makeup artist, wardrobe technician, choir singer, tiler (guardian of the entrance to ensure privacy), marshal, and registrar by their respective caps.

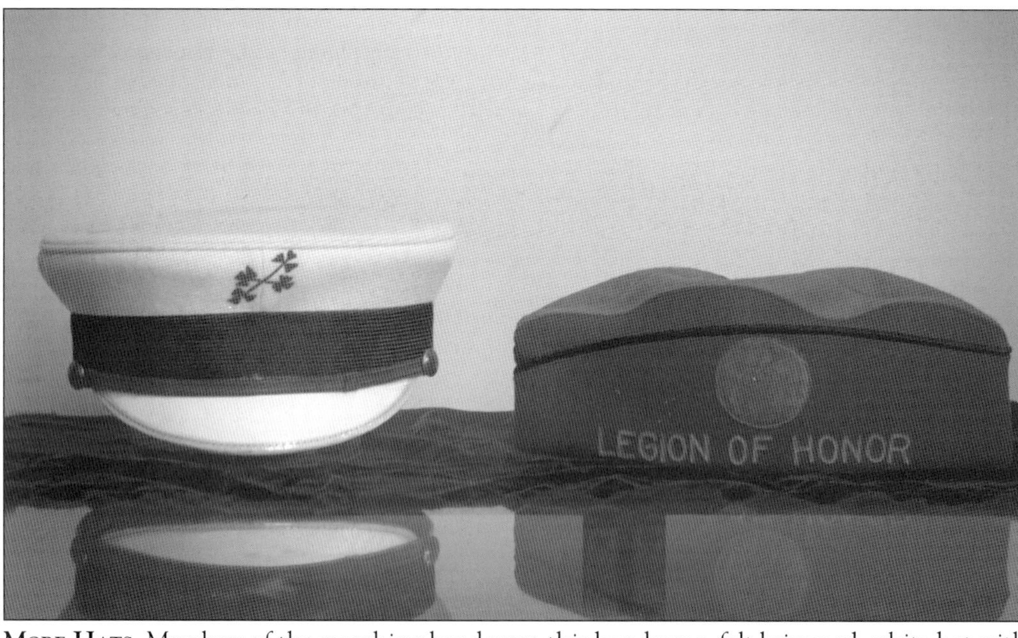

More Hats. Members of the marching band wore this handsome, felt-brimmed, white hat with a gold band as part of their uniform. At least three band members wore this particular hat over the years and signed their names inside it. The signatures of J. Gwin, C. Peterson, and S. Aker identify their usage. The Legion of Honor donned a traditional, military-style black cap, lined with gold trim and a red top and a stamped leather double-headed-eagle patch reading "32."

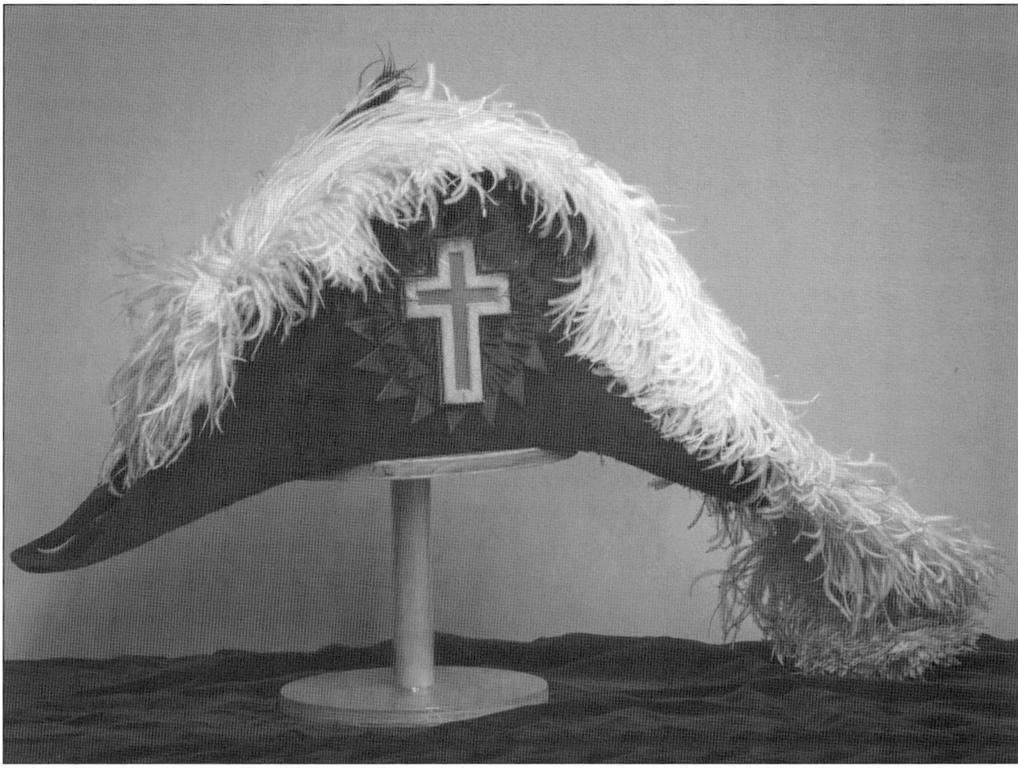

Chapeau of the Order. York Rite Commandry members wear an ancient military-style headpiece with black binding highlighted with an ornate white ostrich plume. This chapeau belonged to S. Reed Megown, and M.G. Lilley and Company of Columbus, Ohio, manufactured it. The matching uniform includes gold-embroidered shoulder straps, gold-colored sleeve crosses, and a sword/scabbard combination.

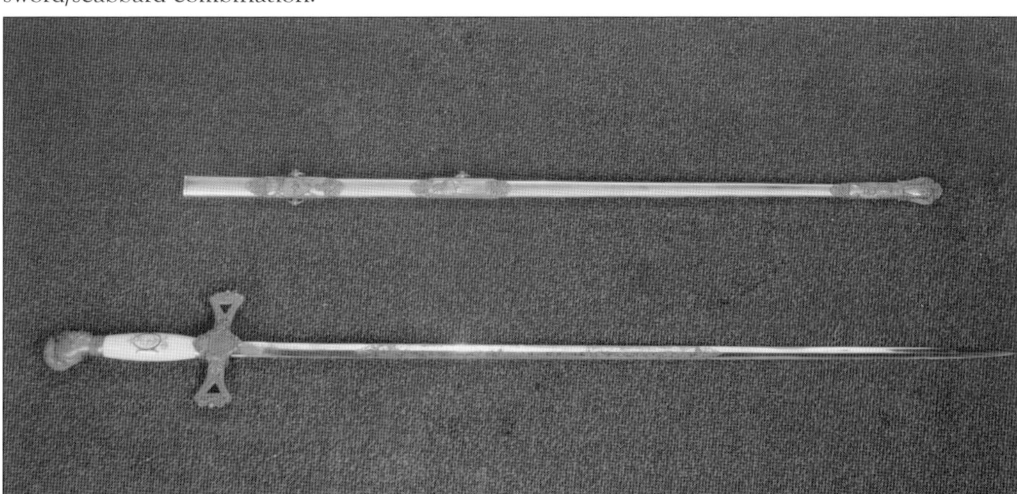

Commandry Sword. Horstmann & Sons of Philadelphia, Pennsylvania, designed this Commandry sword and scabbard for member John J. Donaldson, according to his strict specifications. The hilt displays a cross, the owner's initials, and a knight's head. The scabbard features numerous designs of religious significance.

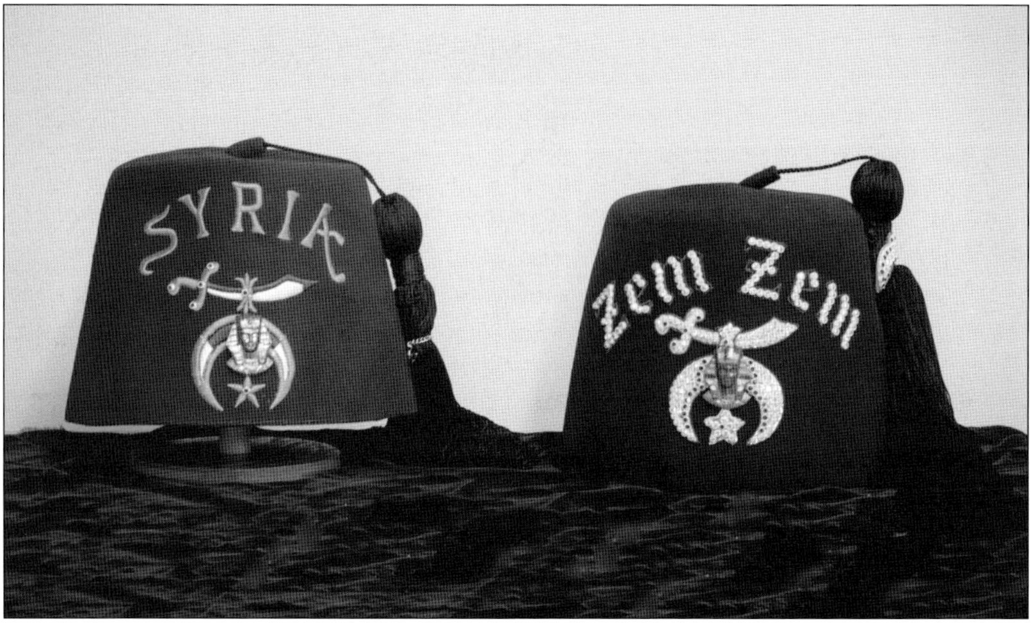

SHRINER'S FEZZES. The fez, named for the holy city Fez, Morocco, where it was first commercially produced, remains the most recognizable of all Masonic hats. The two pictured fezzes represent membership in the Syria Shrine of Pittsburgh and Zem Zem Shrine Center of Erie. Nobles of the Shrine generously support 29 crippled children hospitals and three burn units, providing free medical care across the United States, Mexico, and Canada.

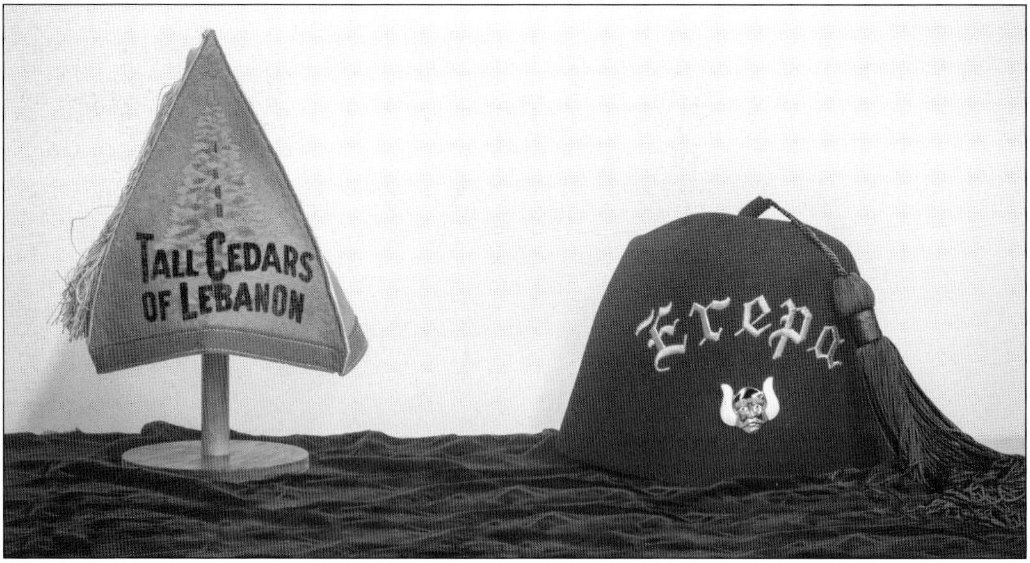

THE TALL CEDARS AND THE GROTTO. The Tall Cedars of Lebanon represents a side degree of Freemasonry with the motto, "Fun, Frolic, & Fellowship." The Tall Cedars charitable arm supports muscular dystrophy and other neuromuscular research. The Mystic Order of Veiled Prophets of the Enchanted Realm (the Grotto) remains open for all Master Masons, providing fun and a fuller life while helping out with dental care for children with special needs. A black fez with a red tassel identifies Grotto members. (Most headpieces shown on pages 73–76 came from the Wallace Library Collection.)

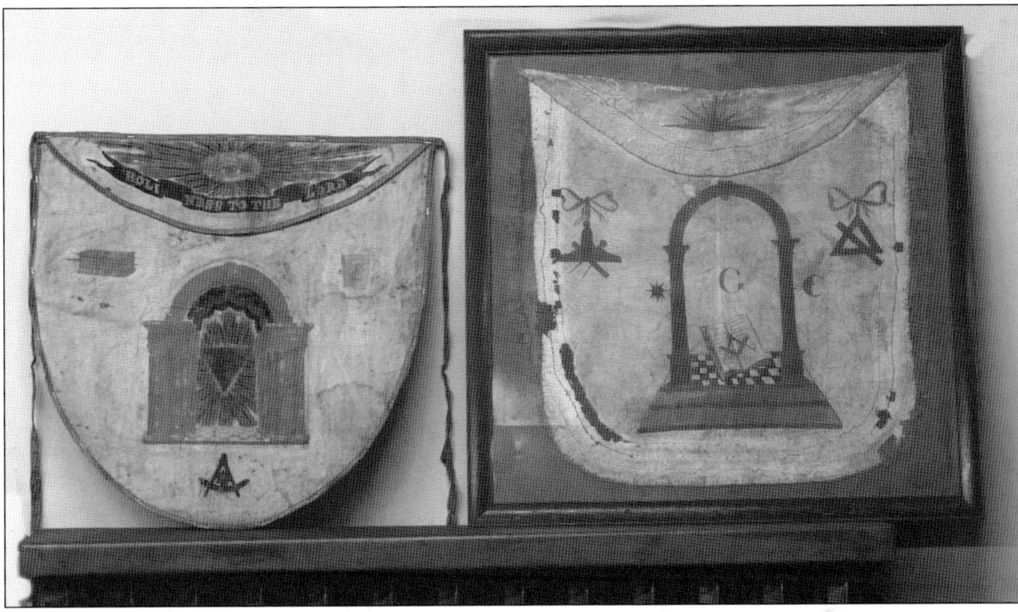

ANTIQUE MASONIC APRONS. These 19th-century aprons date back well over one century. Alex Pearson McCready donated a hand-painted lambskin apron to Mahoning Lodge No. 243. He believed it belonged to his father, George Pearson Sr., born in Darby, Pennsylvania, in 1779, who died in New Castle in 1871. Edward Ward wore a York Rite linen apron in 1822. His descendant, brother Howard Ward, Past Master of Westminster Lodge No. 727, donated it to the Cathedral's Wallace Library.

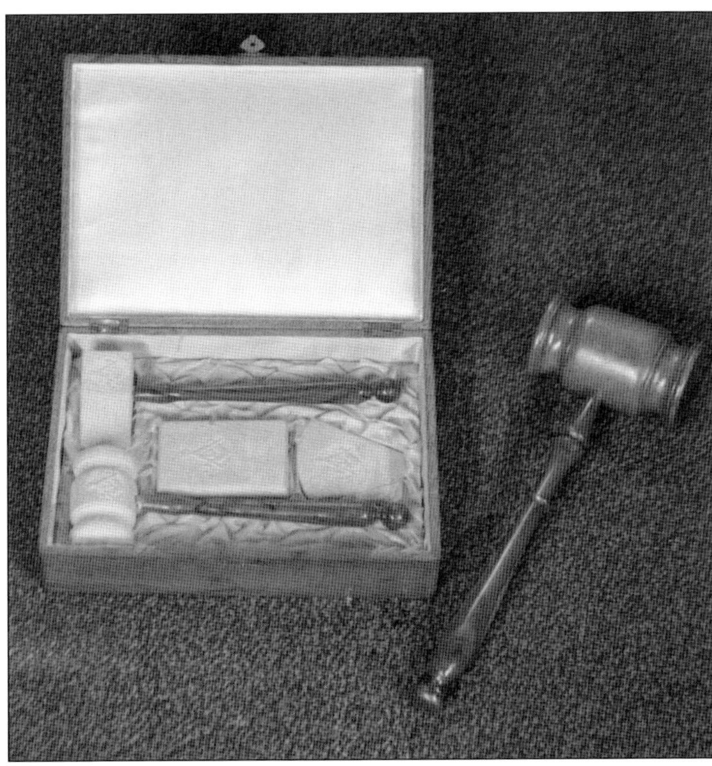

MASONIC GAVELS. Brother William E. Ray received these symbols of authority in 1961. These ornamental gavels and ashlars, manufactured in Jerusalem and embossed with Masonic symbols, used stone from King Solomon's mine.

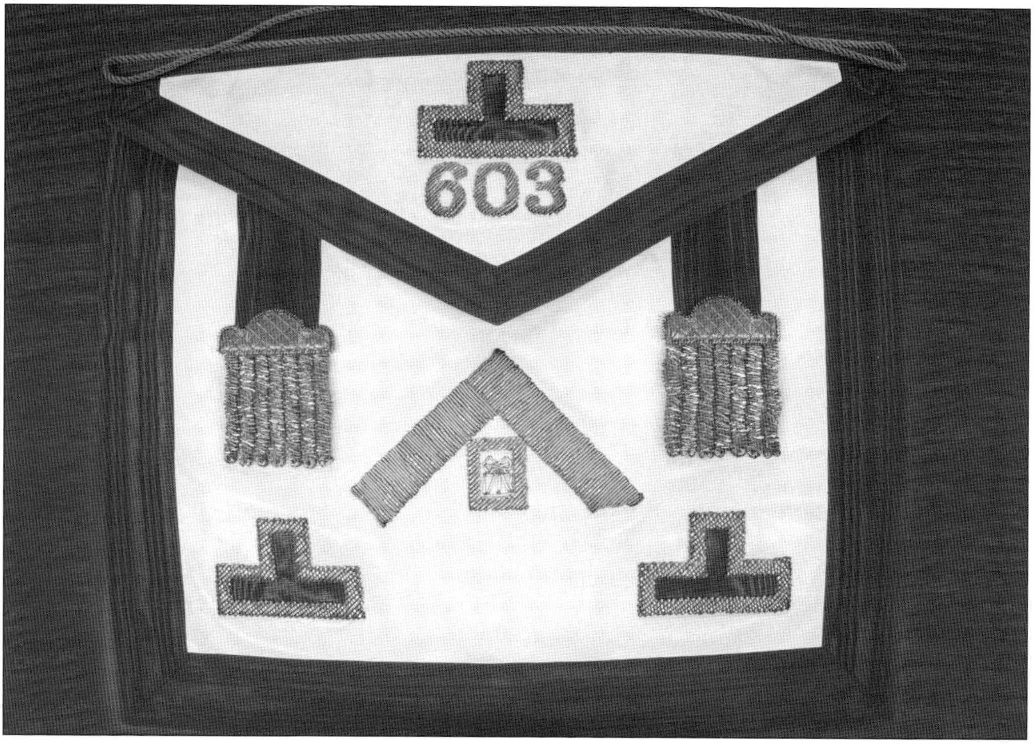

Past Worshipful Master's Apron. This apron presented on February 23, 1960, memorializes the service of Carl C. Kennedy, who served as Worshipful Master of Grove City Lodge No. 603 in 1909. The library received this apron as a gift to commemorate the 50th anniversary of his service.

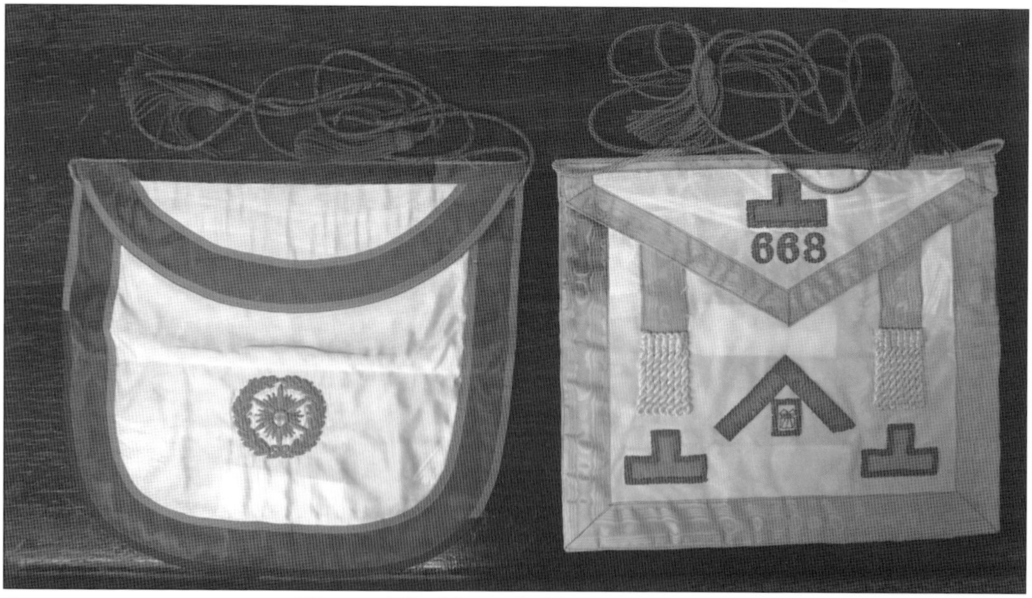

Aprons from around the World. The library holds a variety of antique aprons, including one from a Scottish Blue Lodge worn by Patrick Anderson, a 1919 Past Master's from Shenango Lodge No. 668, and another used by District Deputy Grand Master Mott Robertson (33rd degree) of the Grand Lodge of Pennsylvania.

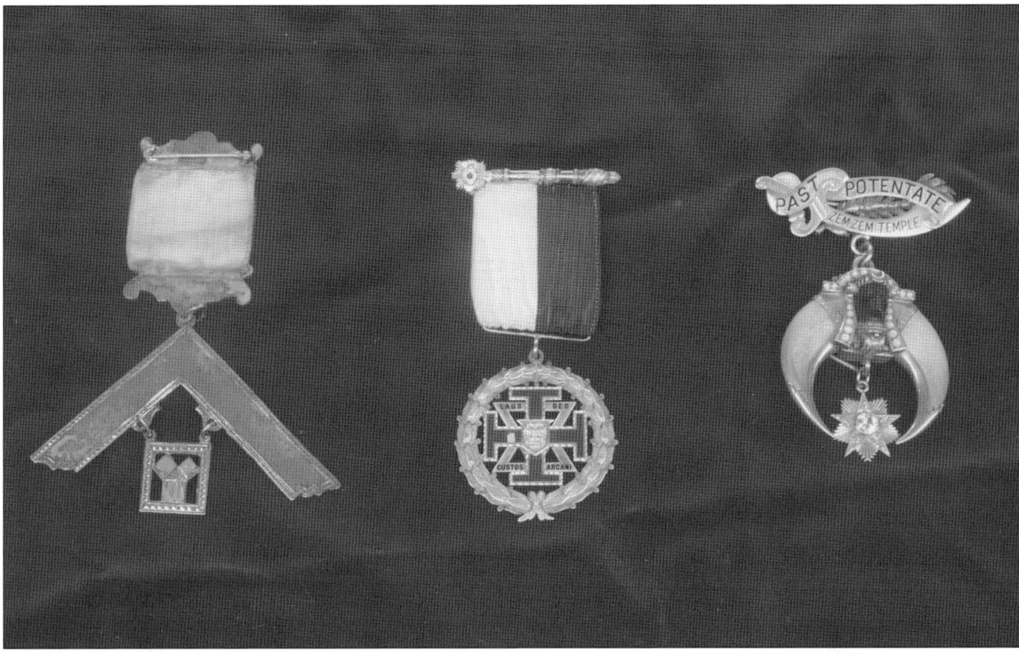

Masonic Jewels. Albert Street of Mahoning Lodge No. 243 received a sterling silver Past Master's jewel in 1944, John Bartholomew earned a Consistory Past Commander in Chief's jewel in 1969, and Levi Taylor merited a Shriners Past Potentate's jewel set with 24 opals from the Zem Zem Temple in 1943.

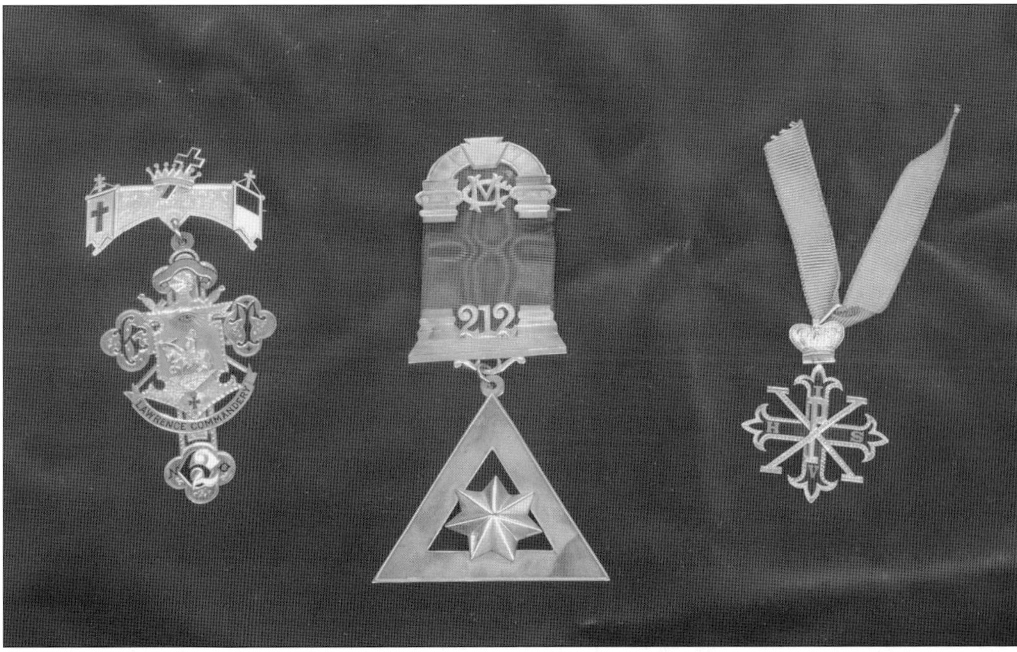

York Rite Jewels. The Wallace Library contains William Butz's 1889 yellow-gold Past Commander's award from Lawrence Commandery No. 62, as well as Levi Taylor's 1926 Past High Priest of the Mound Royal Arch Chapter No. 21 medal and his enamel and 14-karat, yellow-gold Red Cross of Constantine Sovern's ribbon.

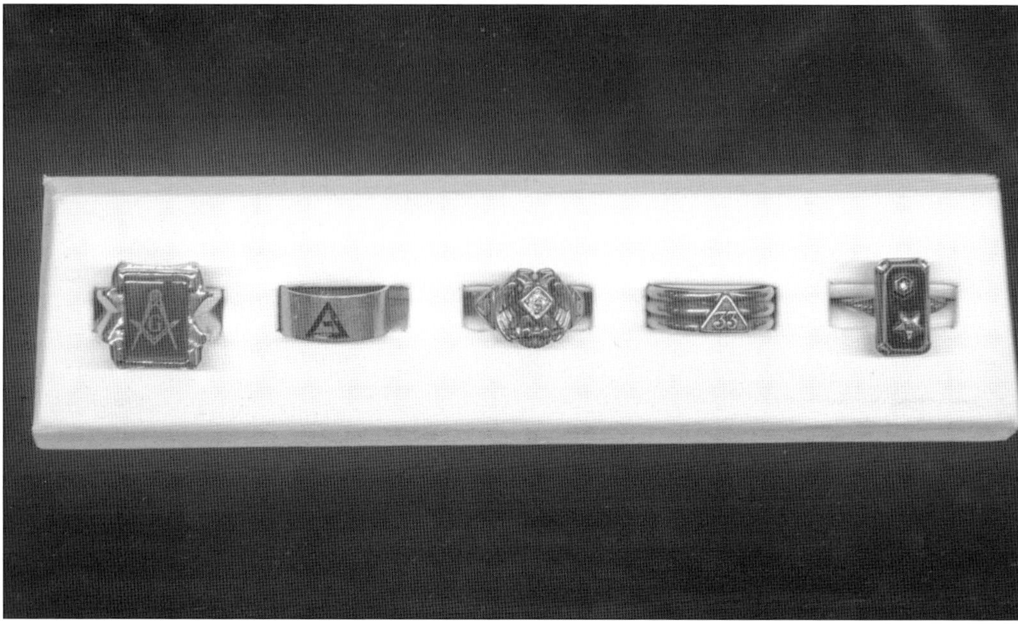

Rings. One can find a variety of rings for every branch of Freemasonry. Represented here are a Blue Lodge, a Lodge of Perfection, a diamond Consistory, and a diamond Shriners ring owned by Rob Cummings, a 33rd degree belonging to Dale Perelman, and an Eastern Star ring.

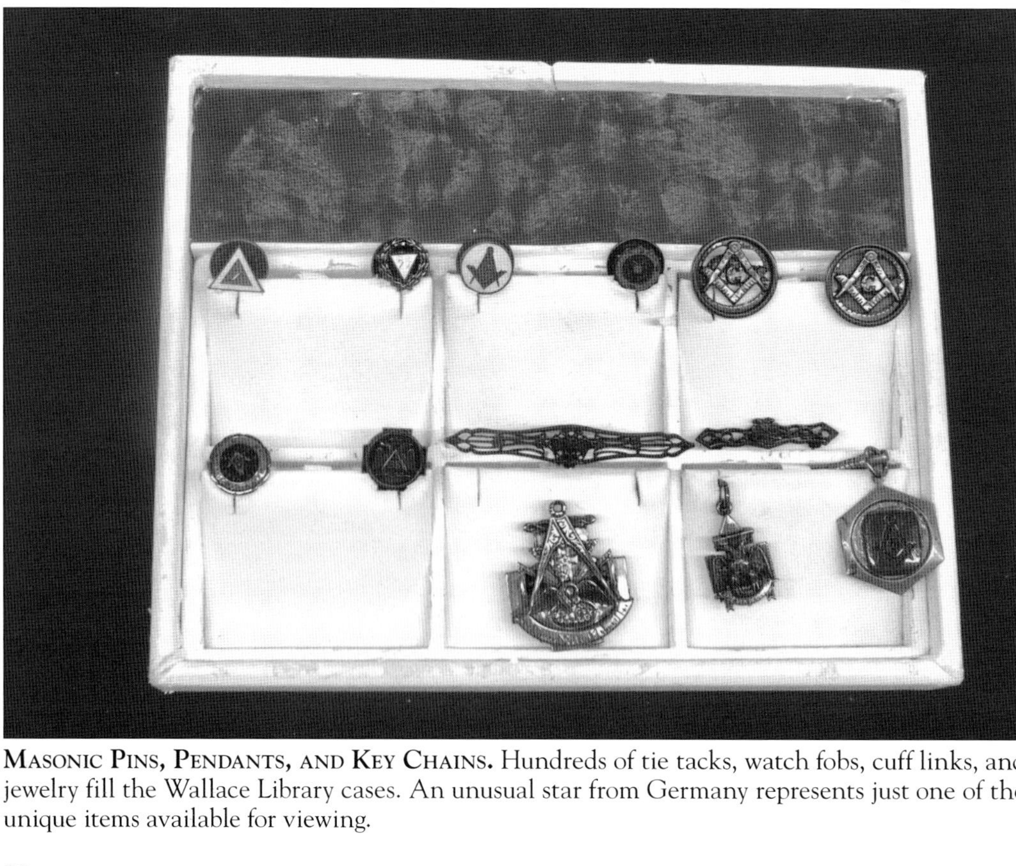

Masonic Pins, Pendants, and Key Chains. Hundreds of tie tacks, watch fobs, cuff links, and jewelry fill the Wallace Library cases. An unusual star from Germany represents just one of the unique items available for viewing.

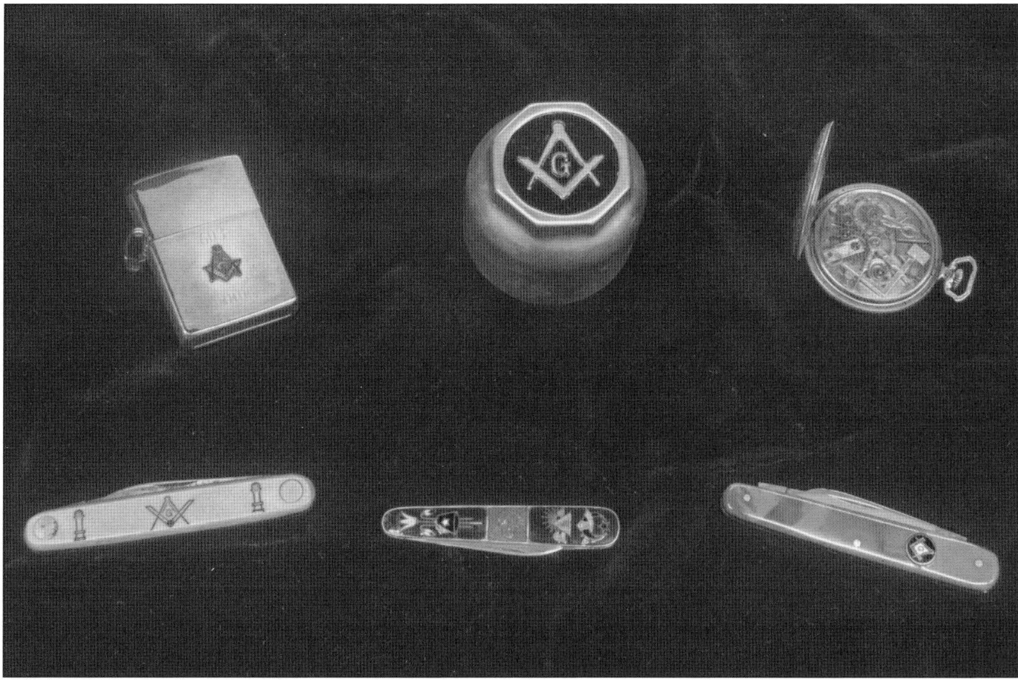

Masonic Regalia. The Wallace Library contains scores of Masonic knives, pocket watches, even a hub cap, a Zippo lighter owned by James L. Cummings of Lodge No. 603, and other paraphernalia, all symbolic of a man's journey through Freemasonry.

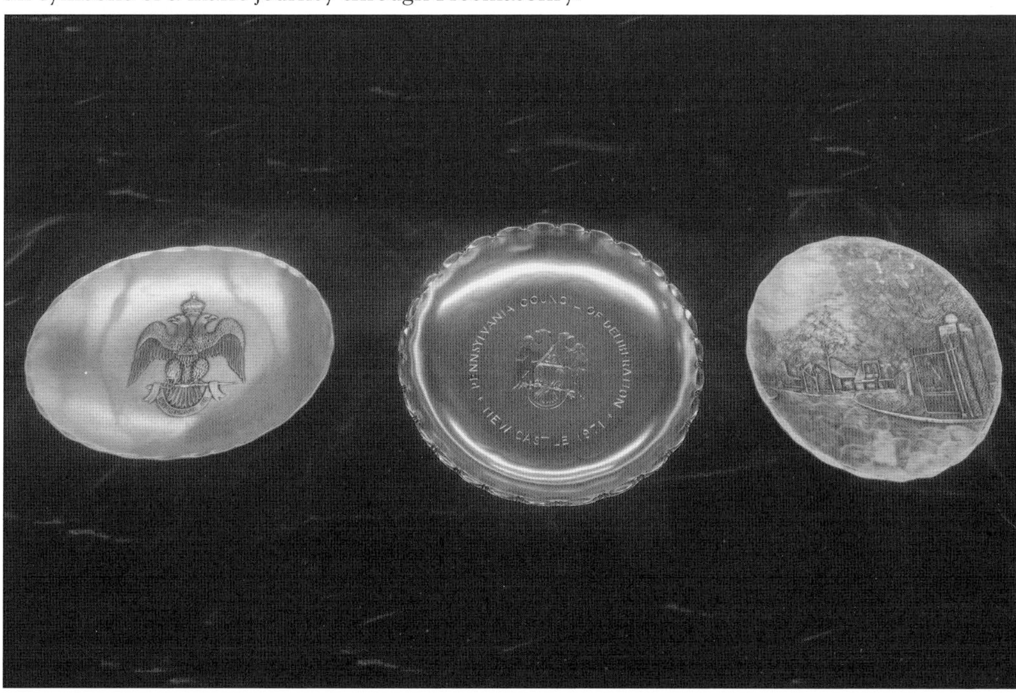

Metal Plates. Wendell August Forge, Arthur Armor, and By-Natale, all from nearby Grove City, produced Masonic-themed, hand-hammered metal plates, some of which can still be purchased today.

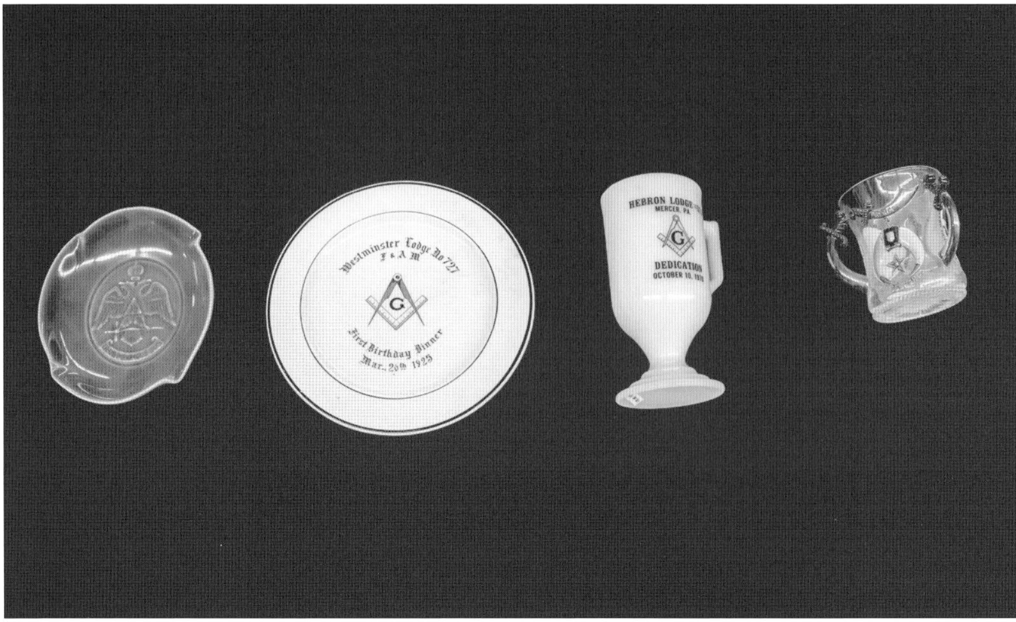

Pottery and Glassware. The Wallace Library includes numerous examples of pottery and glassware. A Rookwood double-eagle from Cincinnati, Ohio, a Westminster Lodge first birthday dinner plate dated March 26, 1925, made by Shenango China of New Castle, a coffee cup marked Hebron Lodge No. 575 of Mercer, Pennsylvania, and a Syria Shrine goblet of Pittsburgh, Pennsylvania, reflect some of the varied items on hand.

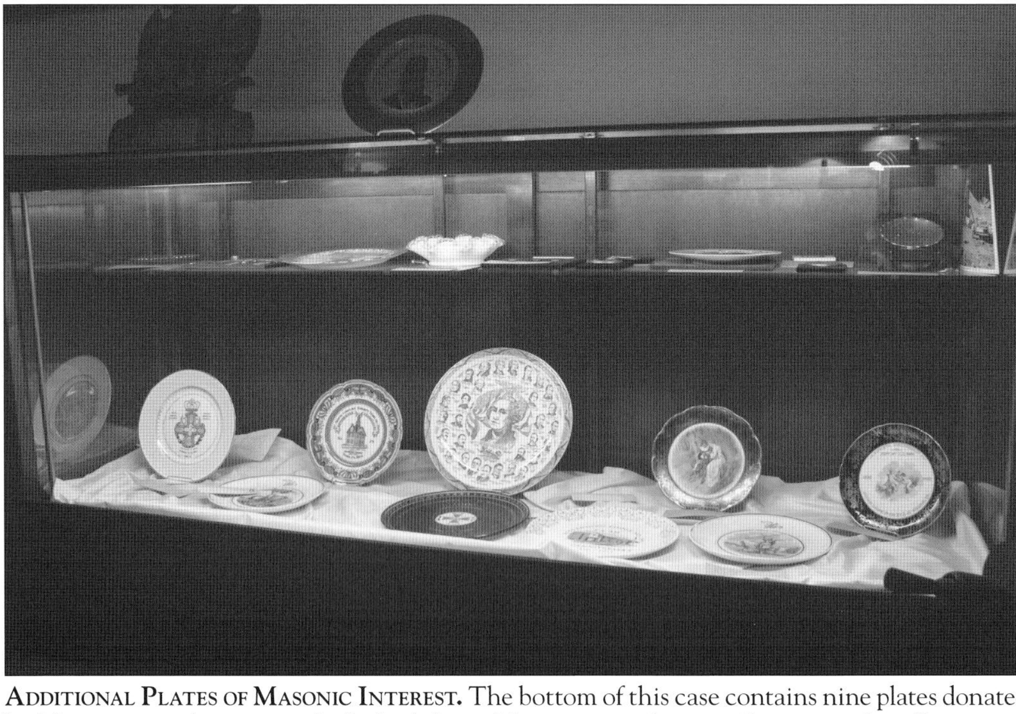

Additional Plates of Masonic Interest. The bottom of this case contains nine plates donated by Masonic members, several of which have Masonic significance. Some merely reflect art of the time.

AUDITORIUM FOLDING SEAT. Scattered throughout the Cathedral, various designs of double-eagle chairs symbolize the Consistory fraternity. The original theatre seats in the auditorium manufactured by the American Seating Company displayed a burgundy leather cushion, an embroidered back panel, and a series of Masonic symbols cast along the side. A wire hanger beneath every seat kept the patron's hat off the floor.

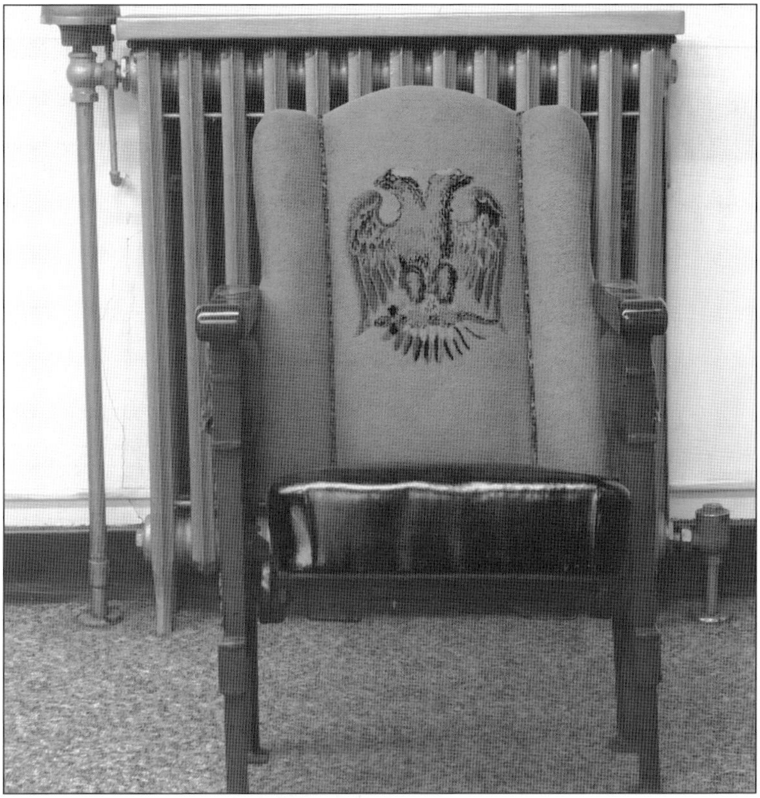

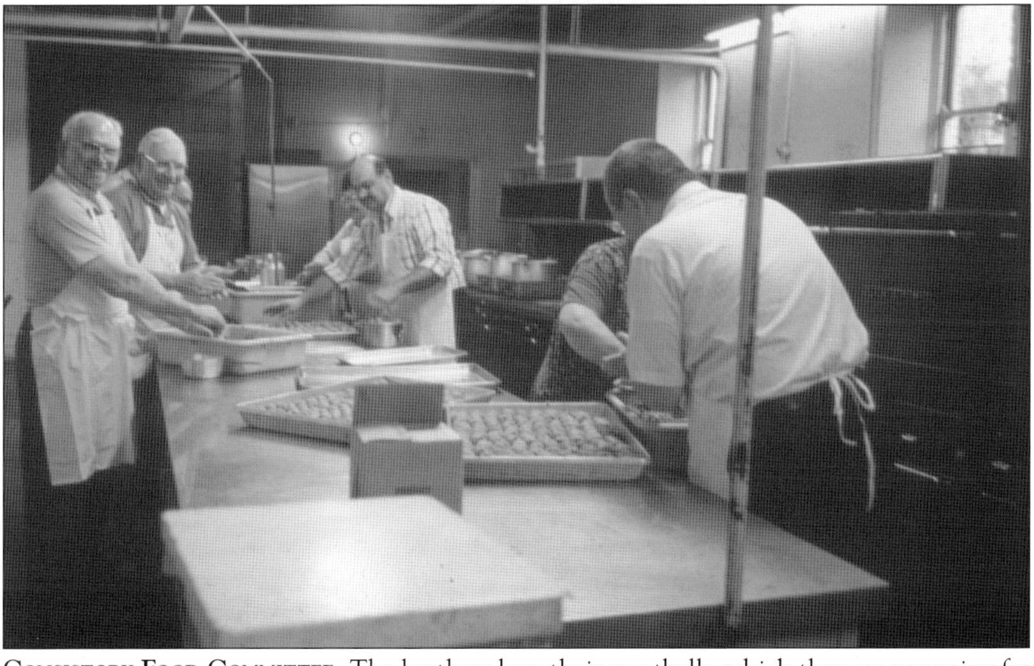

CONSISTORY FOOD COMMITTEE. The brothers love their meatballs, which they are preparing for the annual Lou Rotunno Rigatoni Rally. The fraternity gives all the proceeds from the dinner to a scholarship fund.

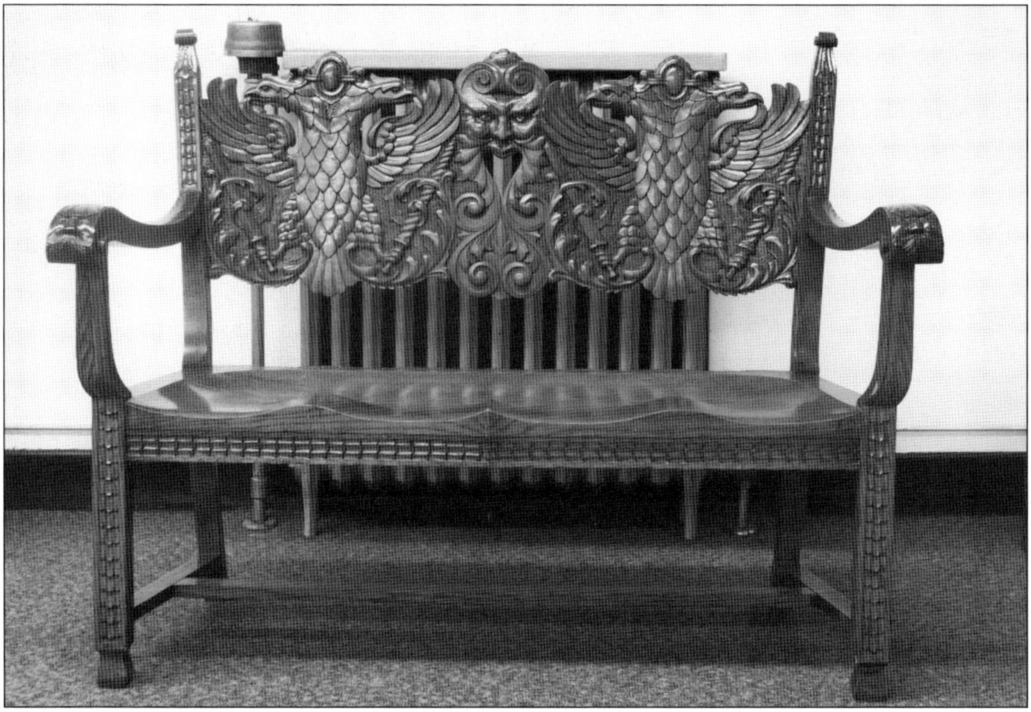

DOUBLE SETTEE. This seat for two includes two sets of double-headed eagles clutching daggers in their talons. A menacing creature in the middle separates the birds. Each armrest ends with a carved snarling face.

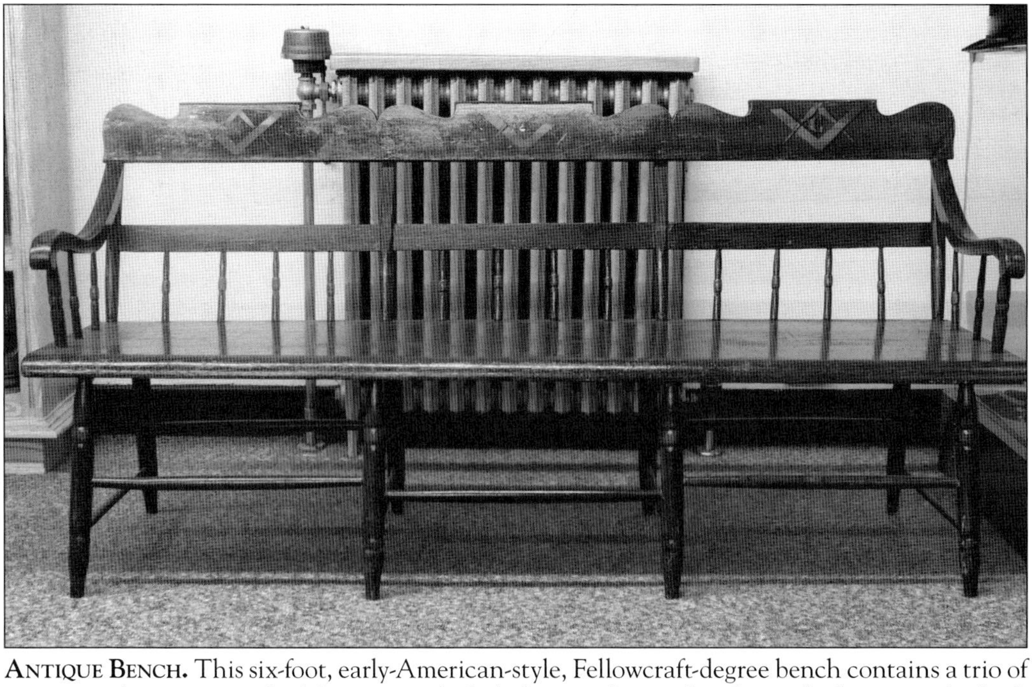

ANTIQUE BENCH. This six-foot, early-American-style, Fellowcraft-degree bench contains a trio of squares and compasses, the Masonic symbol. A design of curved and straight lines ties the look of the piece together.

Green Folding Chairs. At one time, the Cathedral used hundreds of these double-eagle-embossed chairs for overflow crowds, banquets, degree presentations, and wherever needed. For special occasions, white slipcovers added a degree of formality. Today, the Cathedral offers these antique folding chairs for sale at $10 each while supplies last.

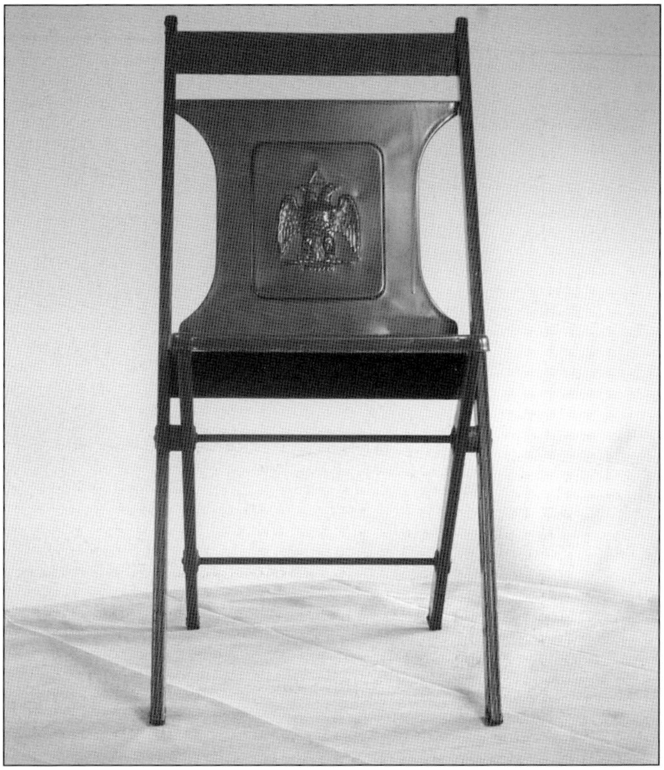

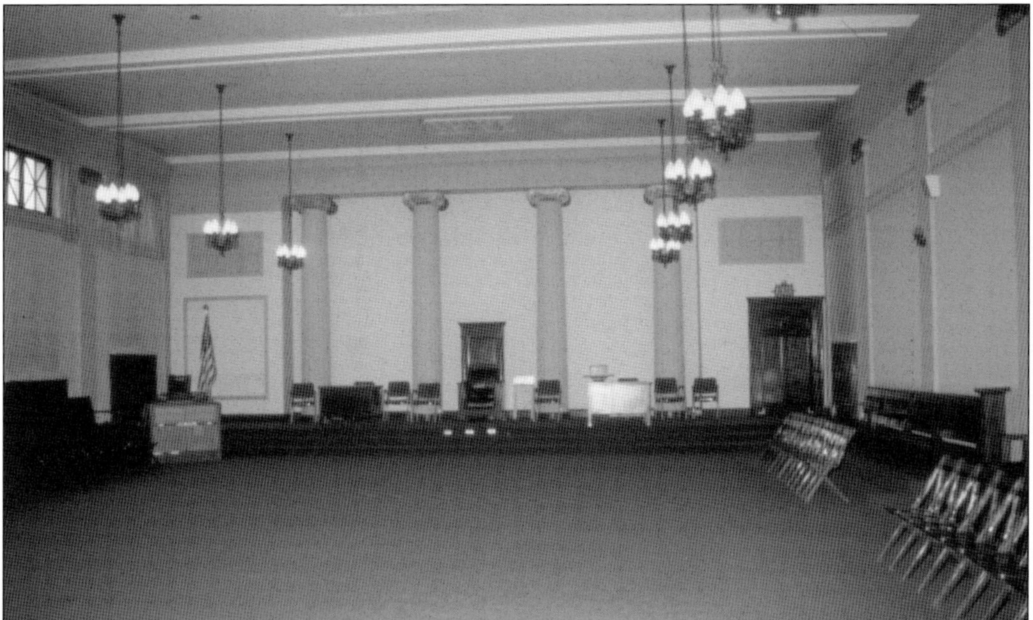

View of East Lodge Room. Two orders of White Shrine of Jerusalem and young men's organizations such as the Builders and DeMolay have used this space in the past. Individuals with questions about Masonry can find most answers to their questions concerning a variety of subjects on the Internet or by contacting www.valleyofnc.org.

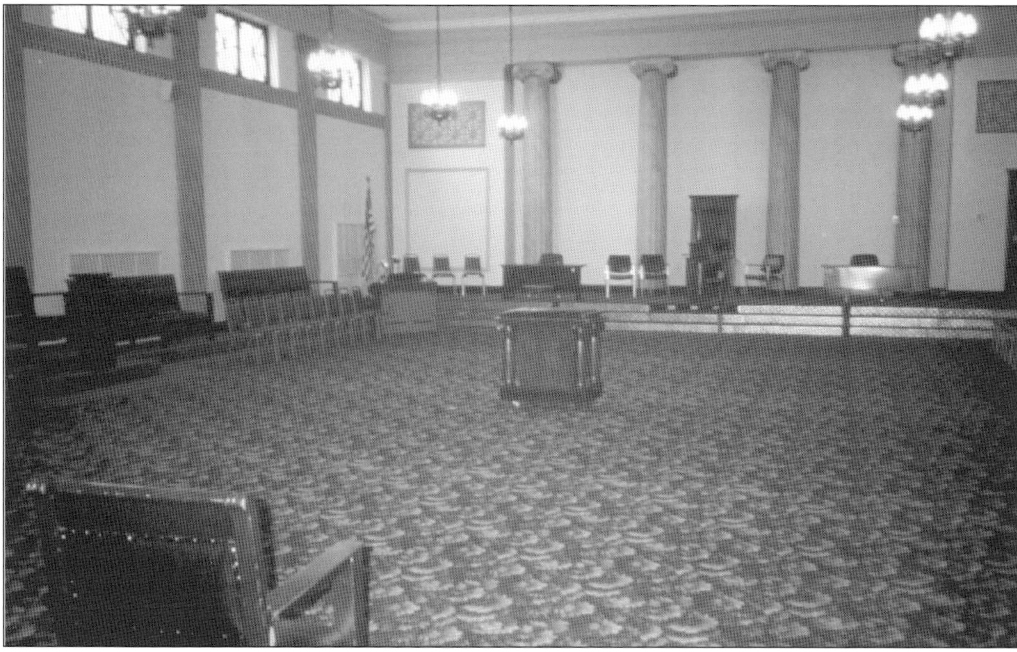

Set Up of East Lodge Room for Consistory. Multiple Masonic organizations use this room for their meetings and rituals. The chairs and altar have been positioned for the Valley of New Castle Scottish Rite Consistory ritual. The Rainbow Girls also meet monthly in this room. The beautifully adorned area may be rented for weddings, small socials, and various ceremonies.

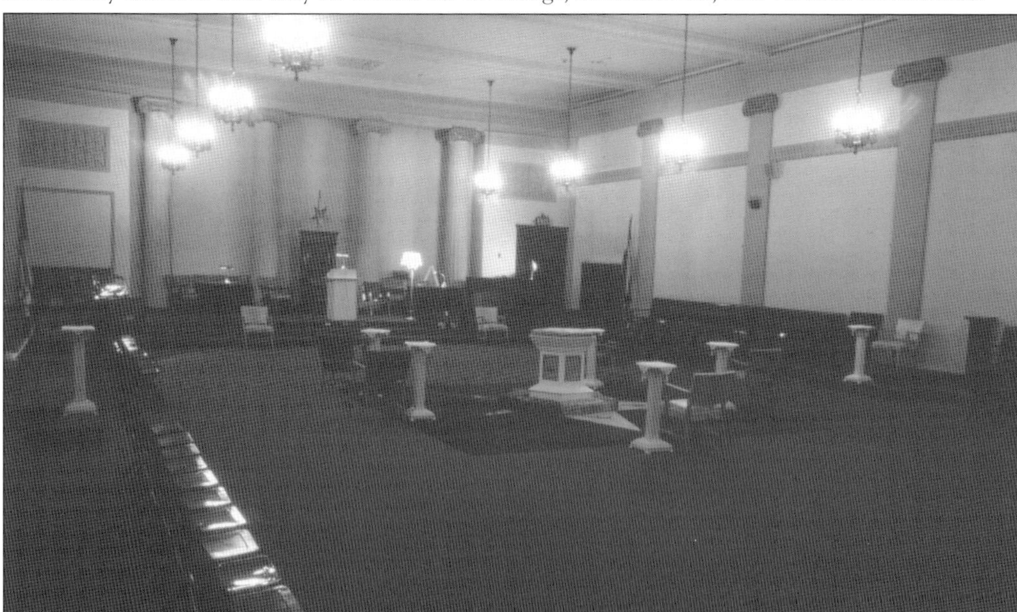

East Lodge Room Set Up for Eastern Star. One of the three Order of the Eastern Star chapters set up this room for their meeting. The women and men of the organization dedicate themselves to the spirit of fraternal love and the desire to work together for good. The star's five rays symbolize lessons learned from the lives of five biblical heroines: Adah, Ruth, Esther, Martha, and Electa.

New Castle Builders Chapter Order of DeMolay Banner. "Dad" Frank S. Land founded DeMolay for 12-to-21-year-old males in 1919, named for Jacques de Molay, the last Grand Master of the Knights Templar, tortured and burned at the stake on March 18, 1314. The organization values seven cardinal virtues: filial love, reverence, courtesy, comradeship, fidelity, cleanness, and patriotism.

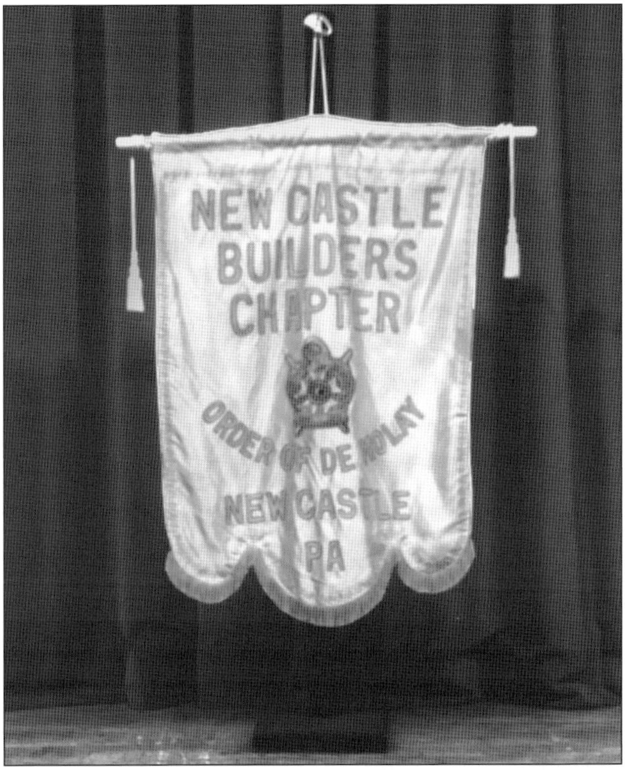

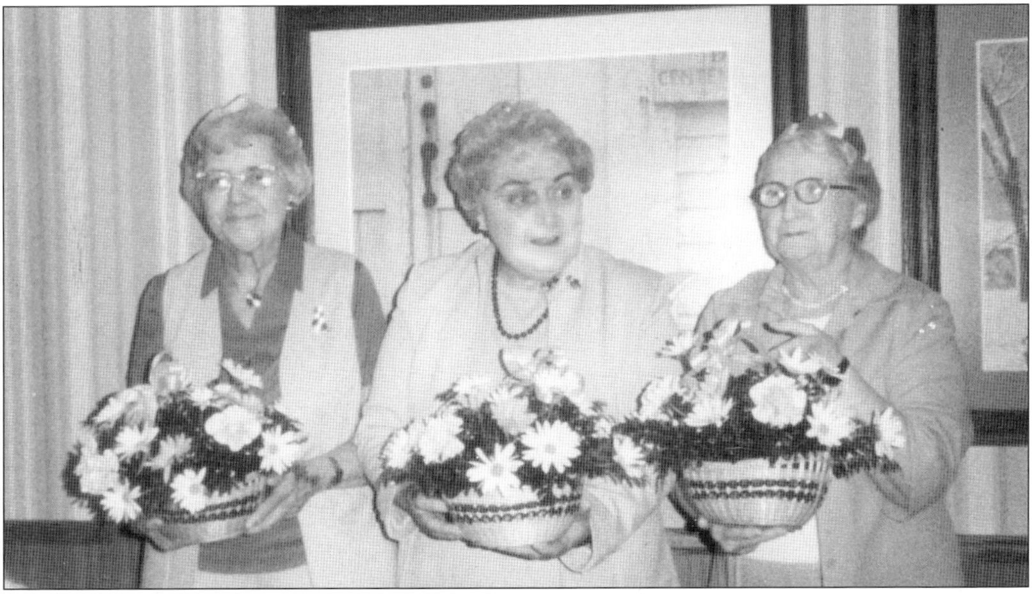

Wimodausis Luncheon Party. These wives, widows, mothers, daughters, and sisters of Master Masons honor Hazel Kerr, Gladys Patterson, and Vivian Walton with flowers. The Wimodausis sorority has built a trust earned through the cooking and serving of meals across decades of banquets, receptions, civic, and Masonic events. The ladies have earned and donated hundreds of thousands of dollars to the Shriners Hospitals, the Children's Dyslexia Learning Center of Western Pennsylvania, the Cathedral Foundation, and a number of local and regional charities.

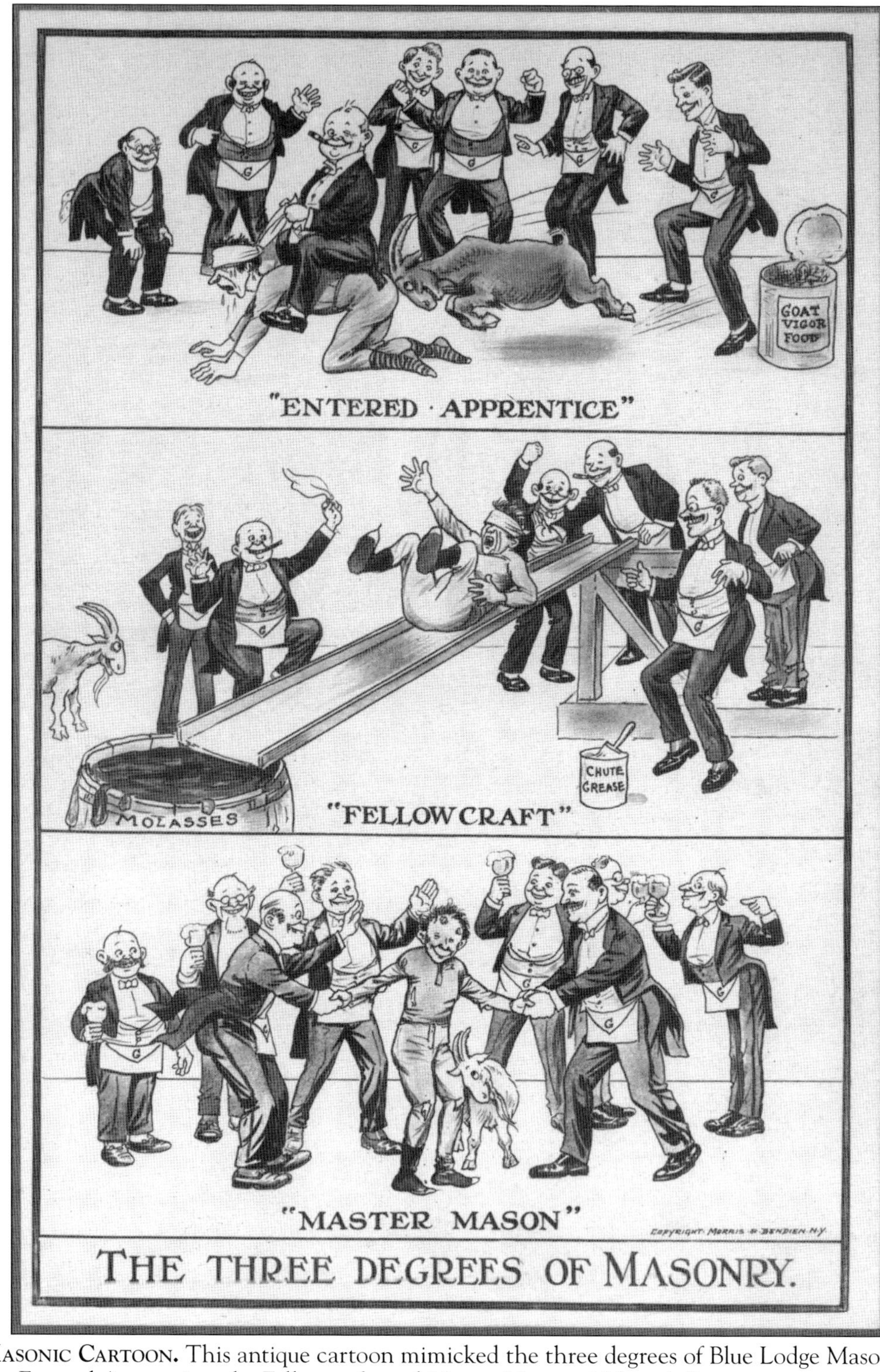

MASONIC CARTOON. This antique cartoon mimicked the three degrees of Blue Lodge Masonry: the Entered Apprentice, the Fellowcraft, and Master Mason. Although Freemasonry represents a serious undertaking, the fraternity offers plenty of opportunities for fun and merriment.

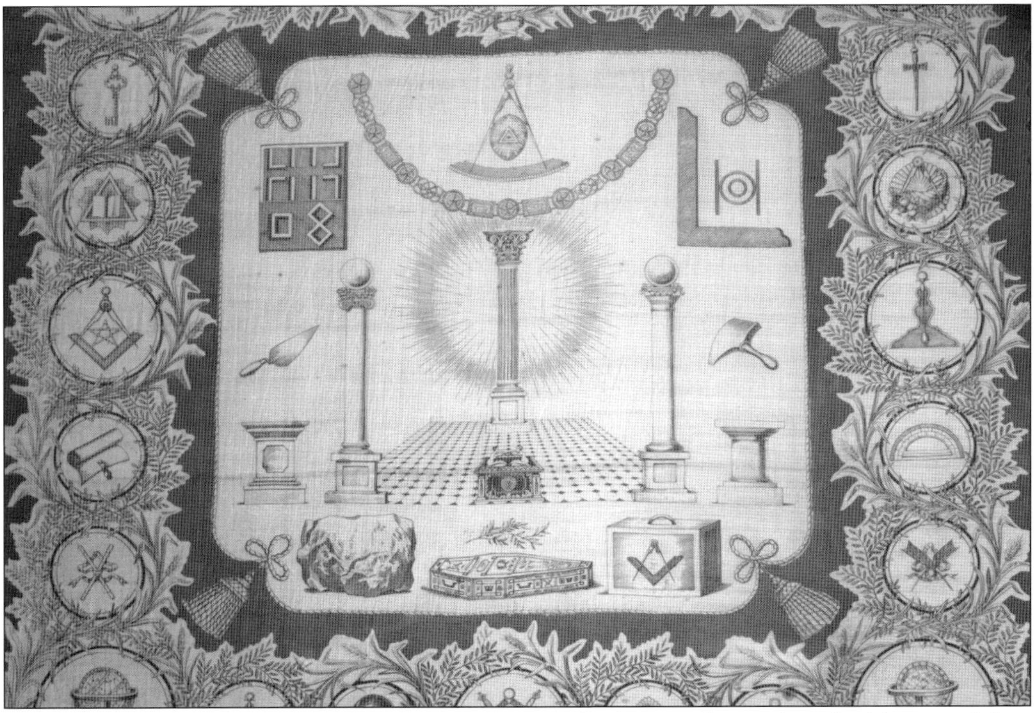

Masonic Wall Tapestry. This decorative antique linen print located in the library exhibits numerous operative Masonic symbols from which speculative Freemasonry grew and thrived.

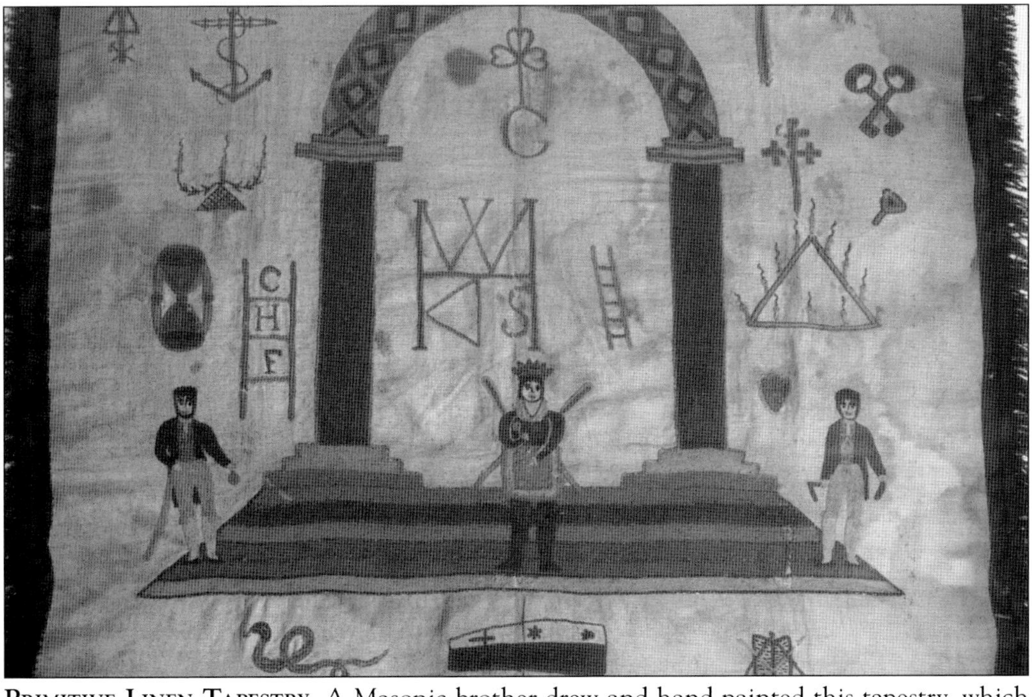

Primitive Linen Tapestry. A Masonic brother drew and hand painted this tapestry, which relates his journey through Freemasonry. He carefully selected those symbols and stories, which he deemed of great importance.

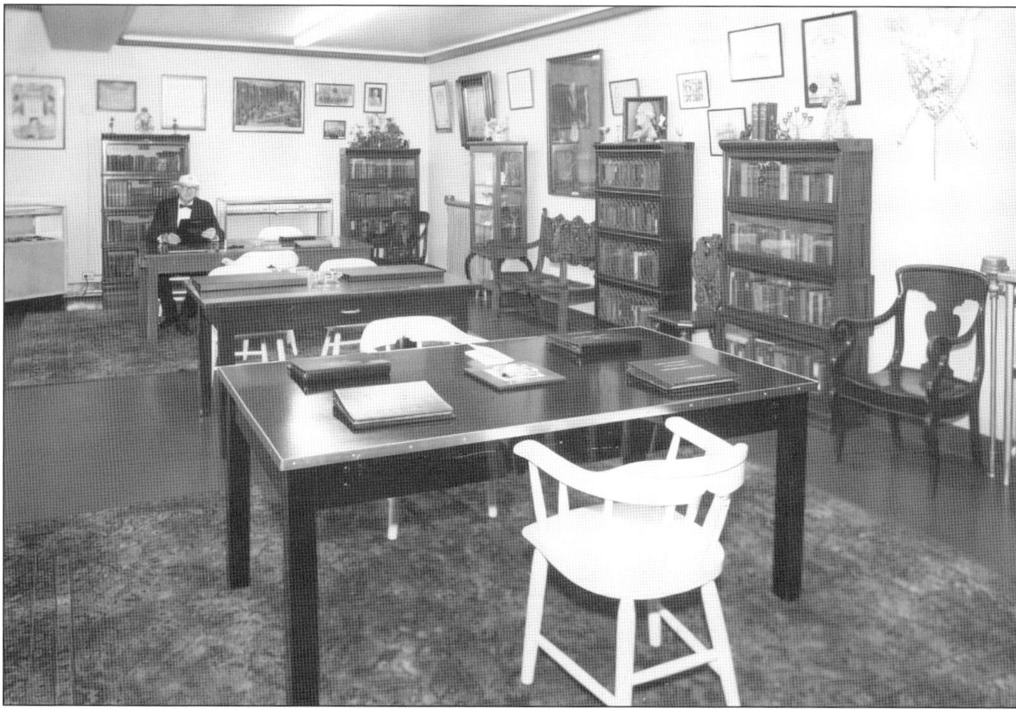

Wallace Library Photograph. A member sits at a table and reads a text. Glass cases hold fraternal artifacts including plates, metals, hats, aprons, patents, and historical documents.

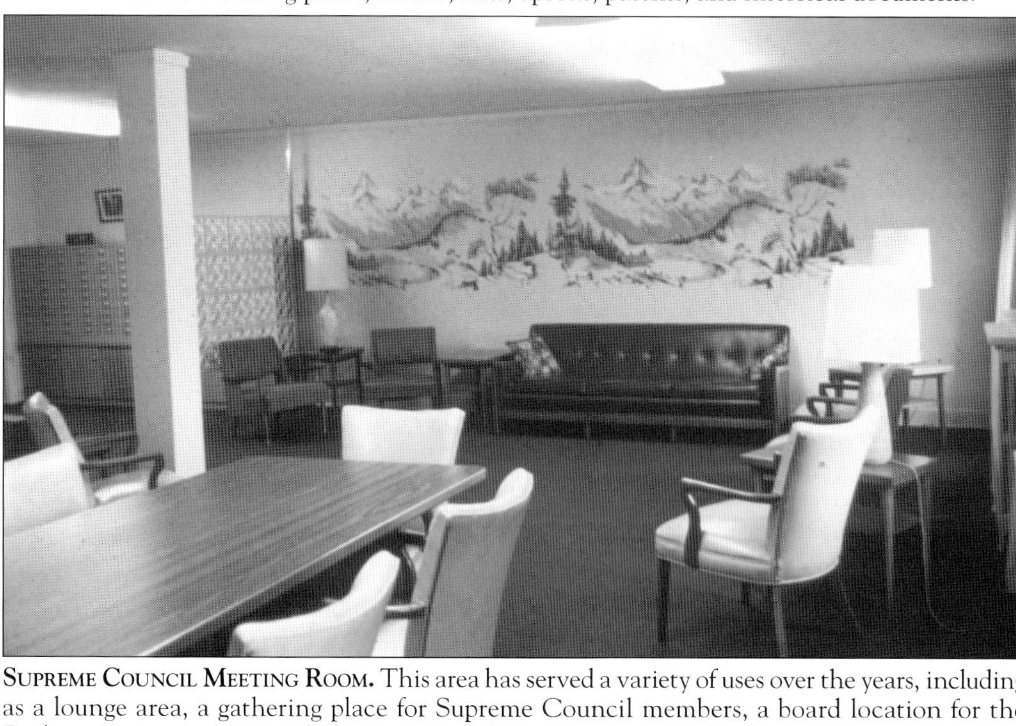

Supreme Council Meeting Room. This area has served a variety of uses over the years, including as a lounge area, a gathering place for Supreme Council members, a board location for the Dyslexia Learning Center of Western Pennsylvania, and a storage facility for the white hats of 33rd-degree honorees.

Five

AUDITORIUM

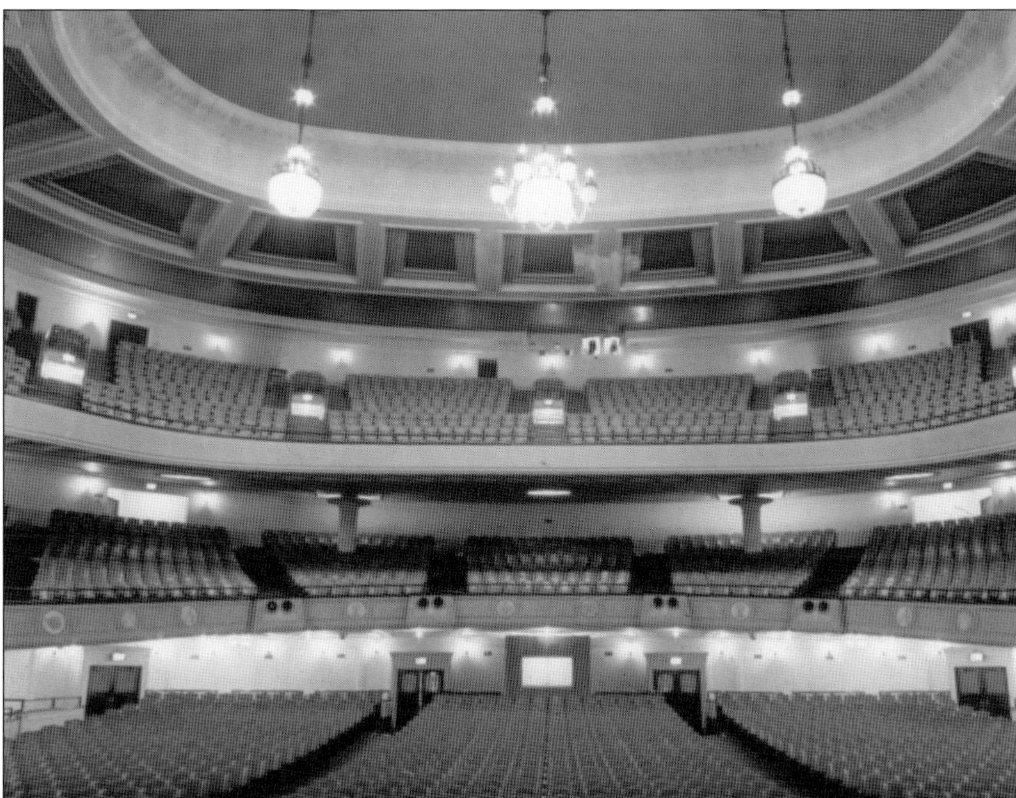

THE AUDITORIUM. The Cathedral auditorium stood as a crowning achievement for the community when completed in 1926, several months after the laying of the cornerstone ceremony. Today, the seating capacity consists of approximately 2,736 seats. (Some of the seats have been placed for cosmetic purposes but are rarely used due to obstructed vision). The building and grounds committee believes the room contains the first cantilevered balcony in Pennsylvania.

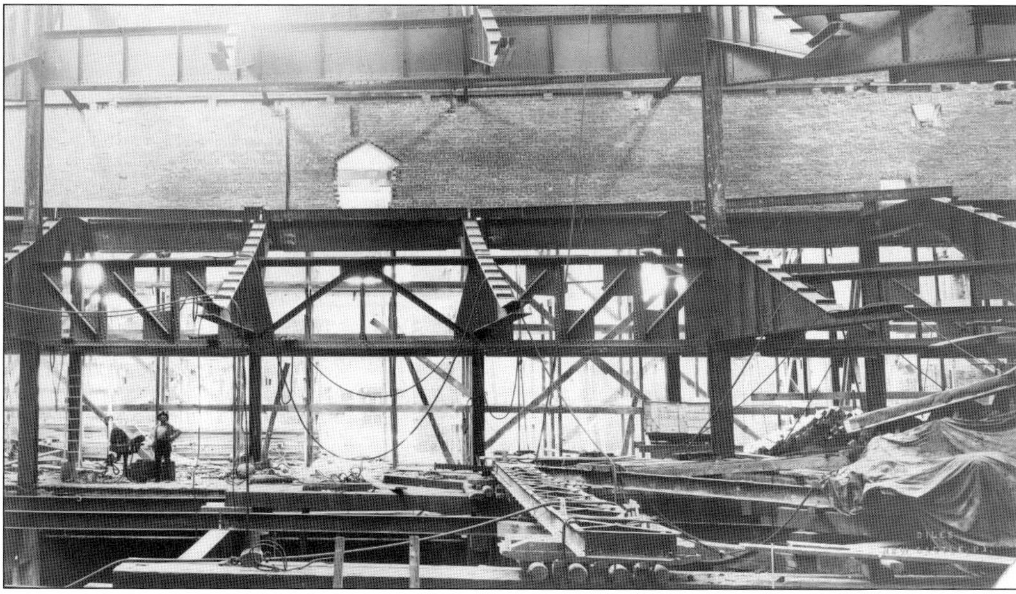

Early Construction. Metal beams and joists form the support for the auditorium, a part of the more than 2.8 million pounds of steel supporting the structure of the Cathedral. A riveter stands by to admire his work. Although invisible to the naked eye in this picture, one of the workmen wrote on a beam in chalk, "Kilroy was here." This August 20, 1925, image by Bile's Photography of New Castle shows the work that continued in the months following the laying of the cornerstone ceremony.

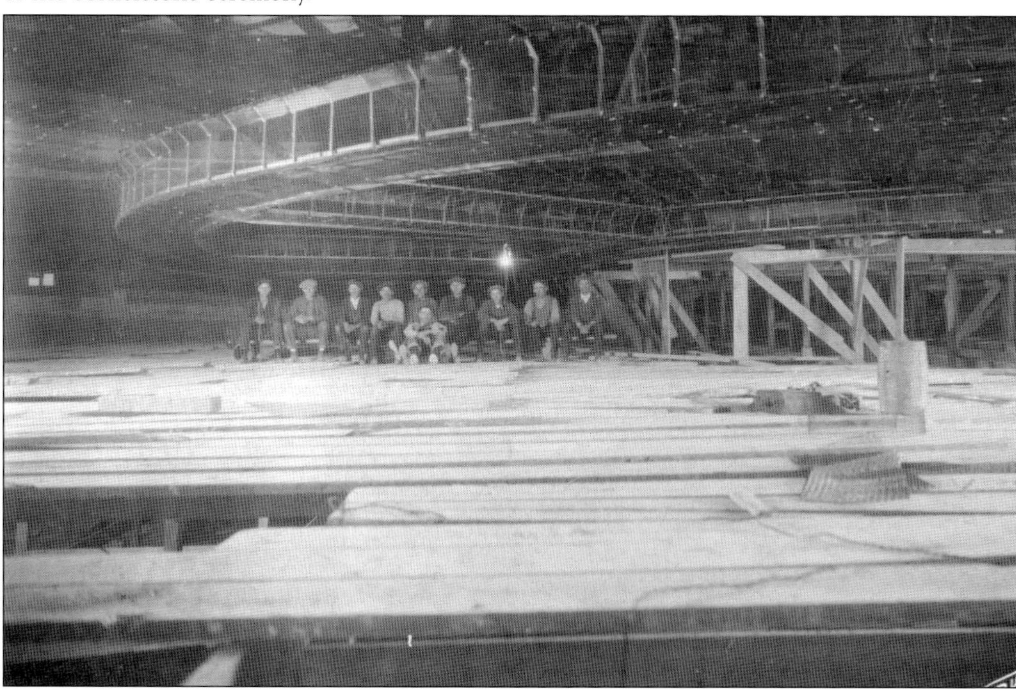

Ceiling Takes Shape. Ten workmen pause for a break and pose for a photograph. The crew sits high above the auditorium floor on temporary scaffolding as they prepare to apply tons of plaster to the ceiling to cover the steel frames and beams.

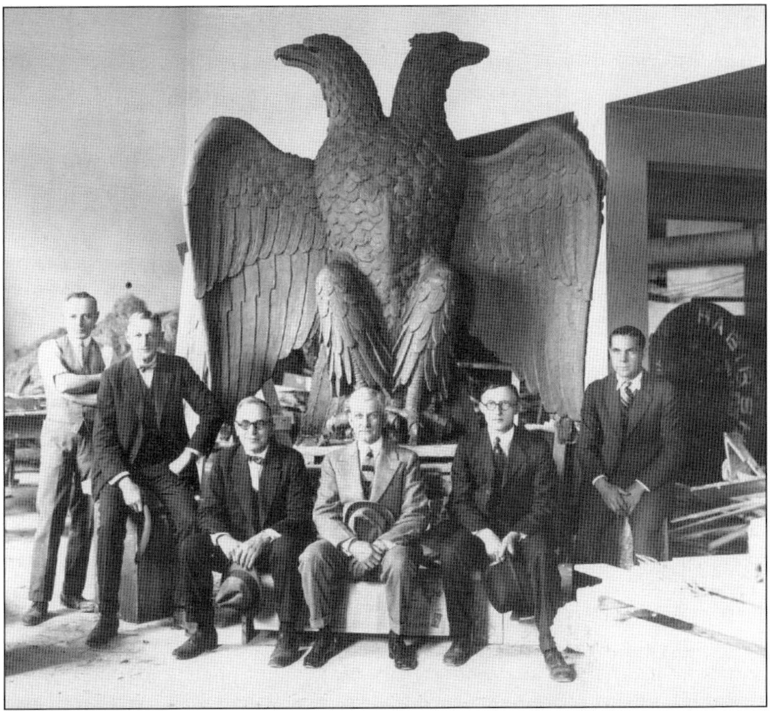

Men with Scottish Rite Double-Headed Eagle. One of the two giant plaster sculptures await positioning atop the stage. One appreciates its size when compared to the six men posed below it. A wooden spool of electrical wire sits in the doorway in the right side of the photograph, a mere fraction of the more than 42 miles required for the construction of the building.

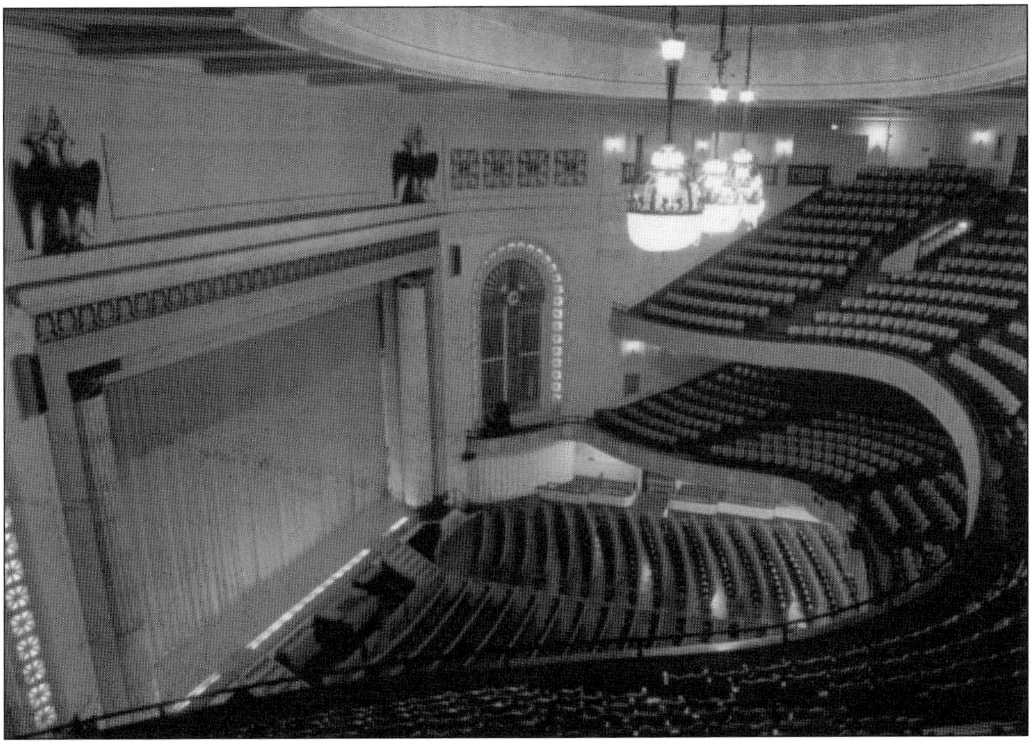

Impressive Auditorium View. Two stately double-headed eagles stand perched above the theater, adding Scottish Rite elegance. A *scaglietti* (imitation marble) pillar borders each side of the stage. Impressive lighting brightens the entire auditorium.

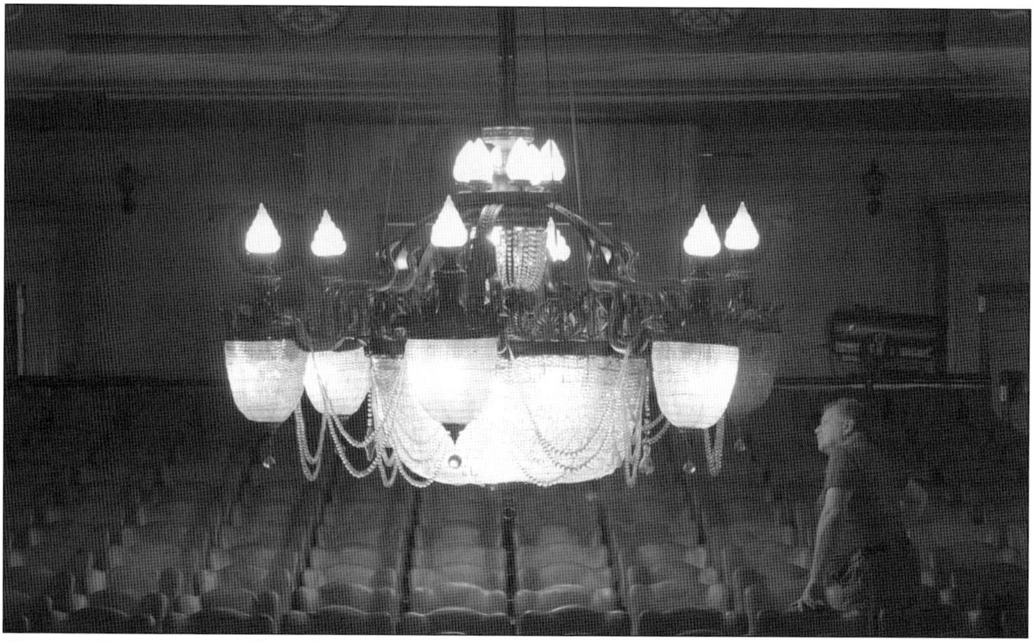

Center Chandelier. The Industrial Appraisal Company values the three auditorium chandeliers at a replacement value of $215,100. The heart of the center unit is identical to the twin fixtures on either side of it. This unit contains six globes connected to the center housing by six pairs of bronze hissing serpents. A torch and a decorative cornucopia design on each globe echoes a similar theme found throughout the fixtures. Chains of crystal beads connect one globe to the next, while crystal finials dangle from the bottom of the piece.

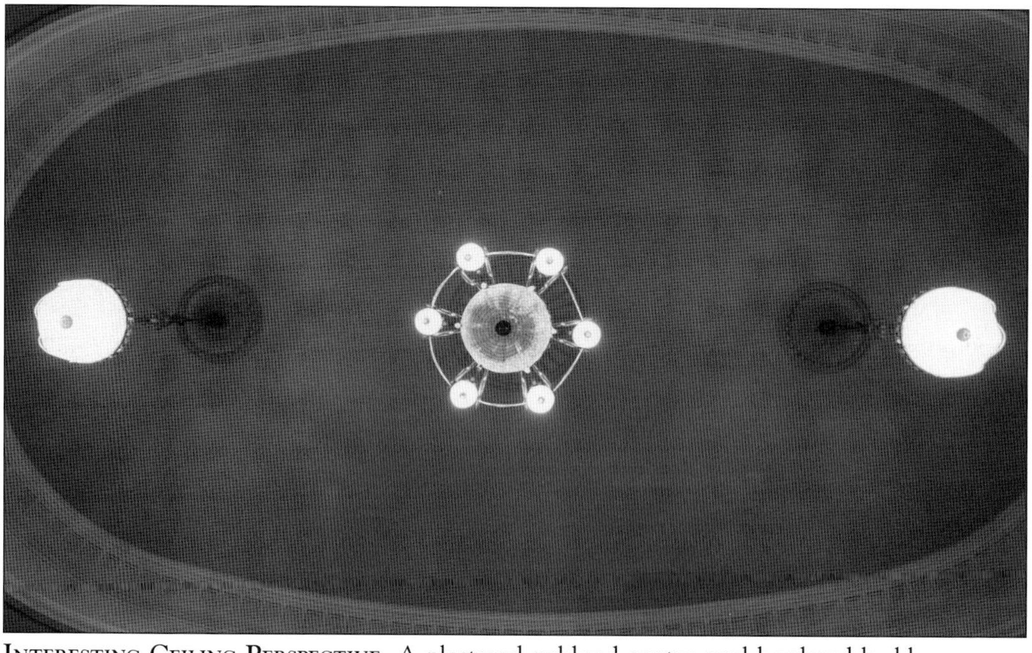

Interesting Ceiling Perspective. A plastered oxblood center oval bordered by blue squares houses the three massive chandeliers. This view provides a unique way of looking at these lighting fixtures.

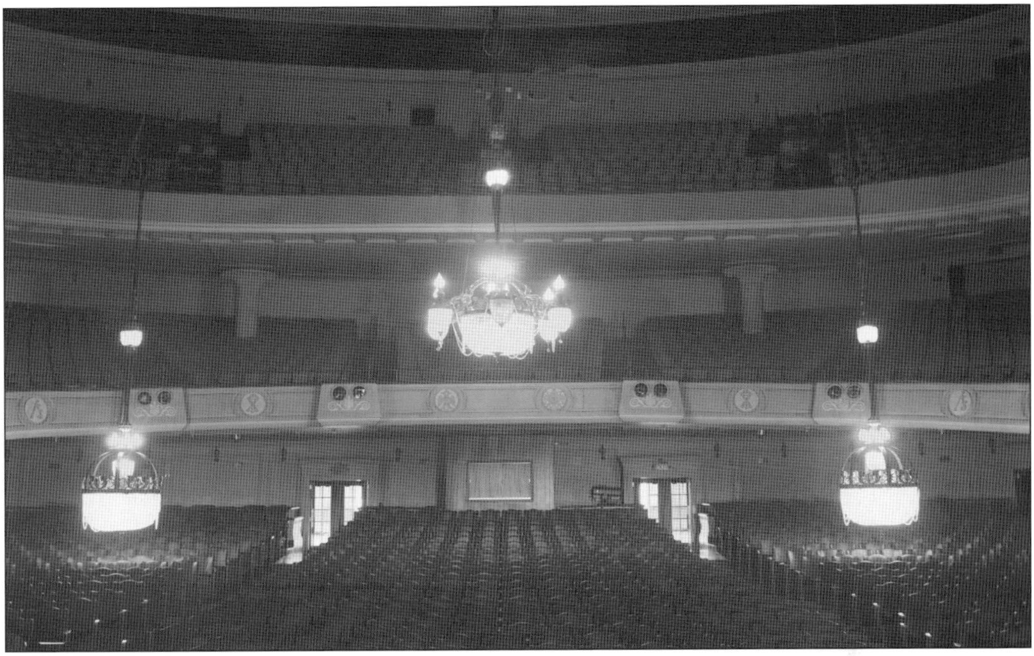

Lowering of the Chandeliers. Hours into the process, with a person feeding electrical wire and his partner cranking the wench, the lights have partially dropped. A third helper coordinates the process from ground level. This dusty and dirty job must be done every few years or as the need arises.

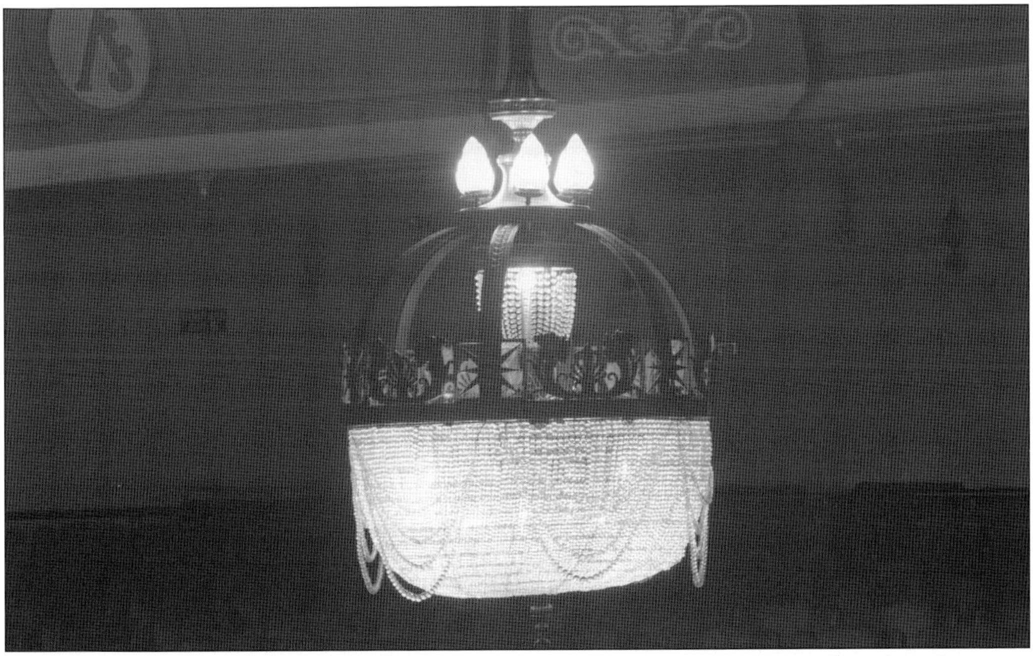

Lowered Chandelier. The Cathedral maintenance and stage crew spent nearly an hour lowering this impressive 2,300-pound fixture to clean the crystals and change from incandescent to LED bulbs. Dropping the light demands a great deal of care due to the age, value, and fragile nature of the piece. Each revolution of a manual hand crank drops the unit by just one inch.

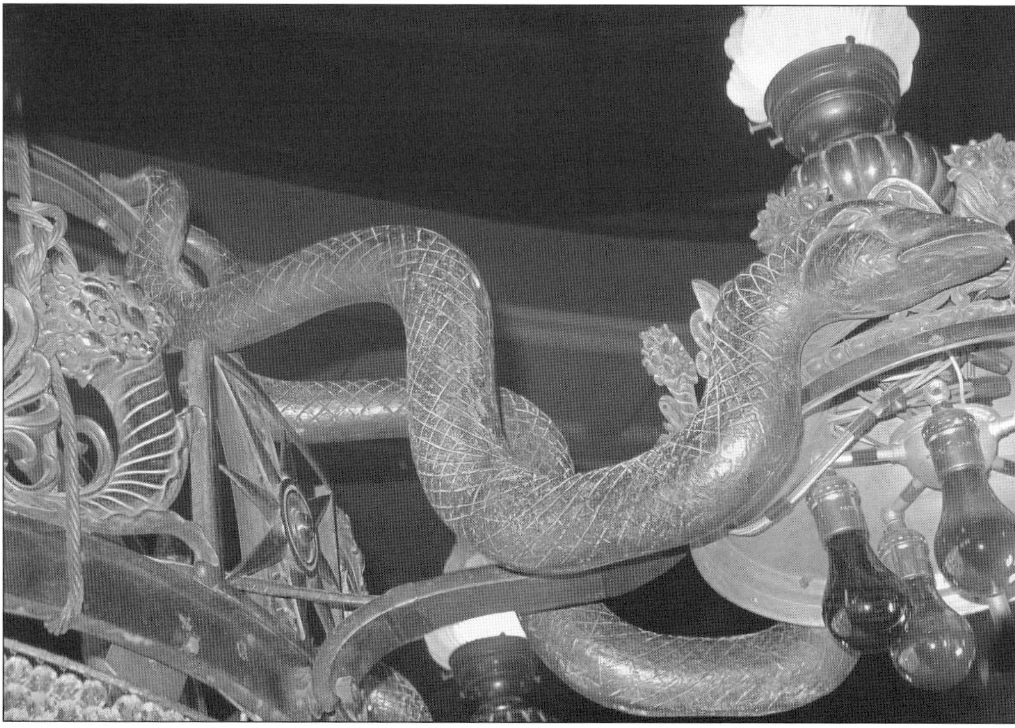

Interior of Center Chandelier. While changing bulbs to modern LED lighting to save on power consumption, photographer Bill Conkle captures a close-up of serpents wrapping their bodies around the structure of the chandelier. Many believe these snakes symbolize the tempter in the garden of Eden. A blue and multicolored globe of the world and views of the sun, moon, and the stars rotating around the sphere represent a hint of paradise to come.

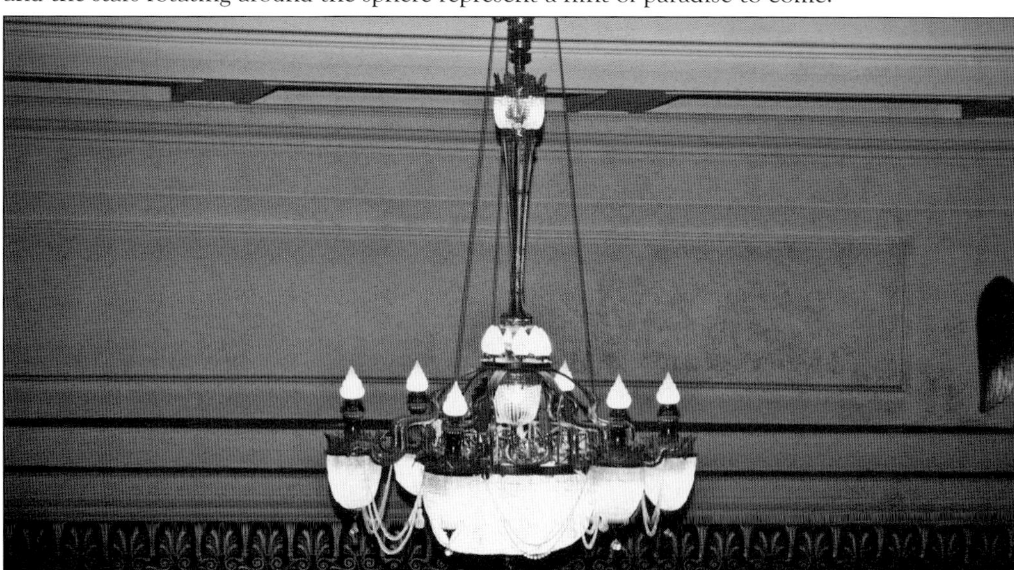

Center Chandelier Detail. High atop the unit sits a blue and multicolored globe of the world. A ring containing views of the sun, the moon, and stars rotates around the sphere. On each side of the 2,300-pound center chandelier hangs a smaller 1,500-pound fixture.

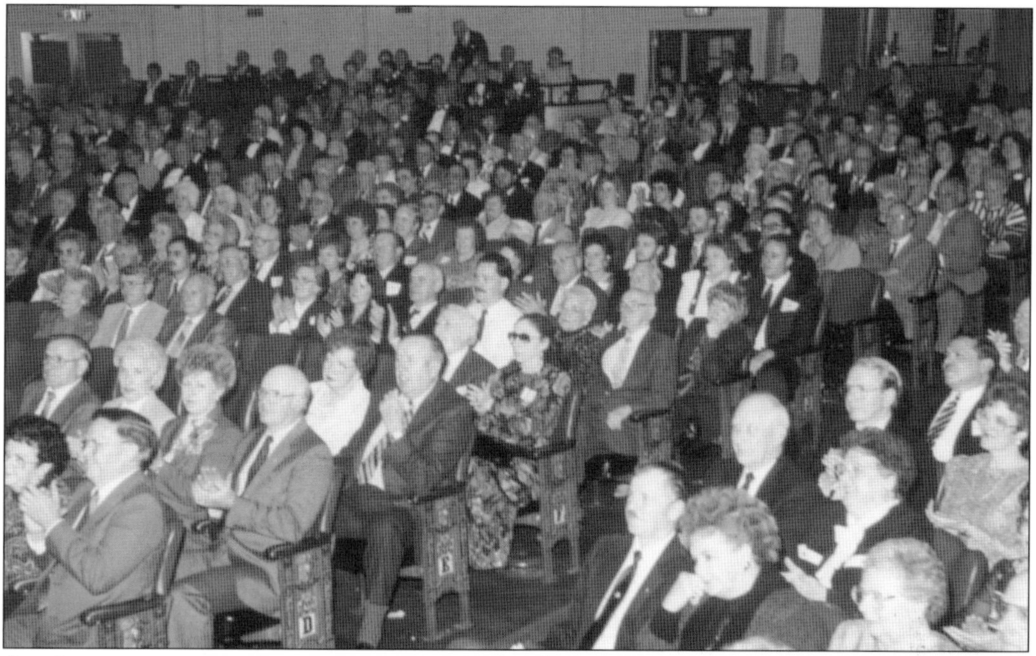

An Audience in the Auditorium. Hundreds fill the ground level of the Cathedral awaiting a performance. The third-row aisle seat on the far right holds Bill McCarrier, a Consistory Past Commander in Chief and an Emeritus Active of the Scottish Rite Supreme Council. Bill's wife, Irene, sits beside him.

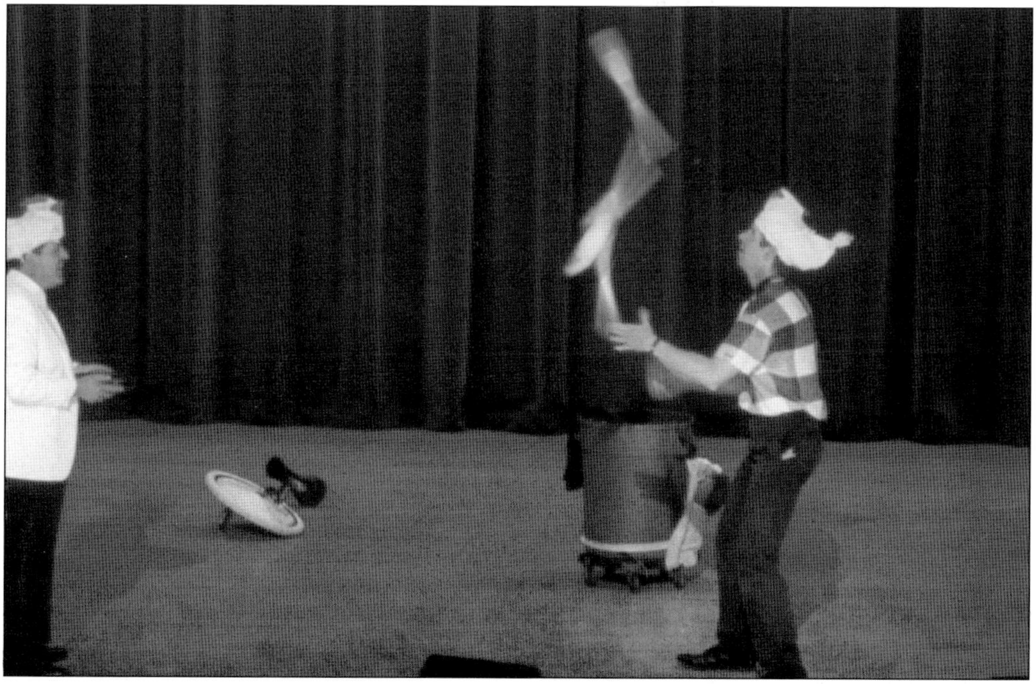

Valley of New Castle Family Night. Commander in Chief Bob Jones cheers on a professional juggler to a packed audience in the 1970s. "Uncle Ed" Shaughnessy, a KDKA legend and celebrity, emceed this act and several others during an evening of fun and entertainment.

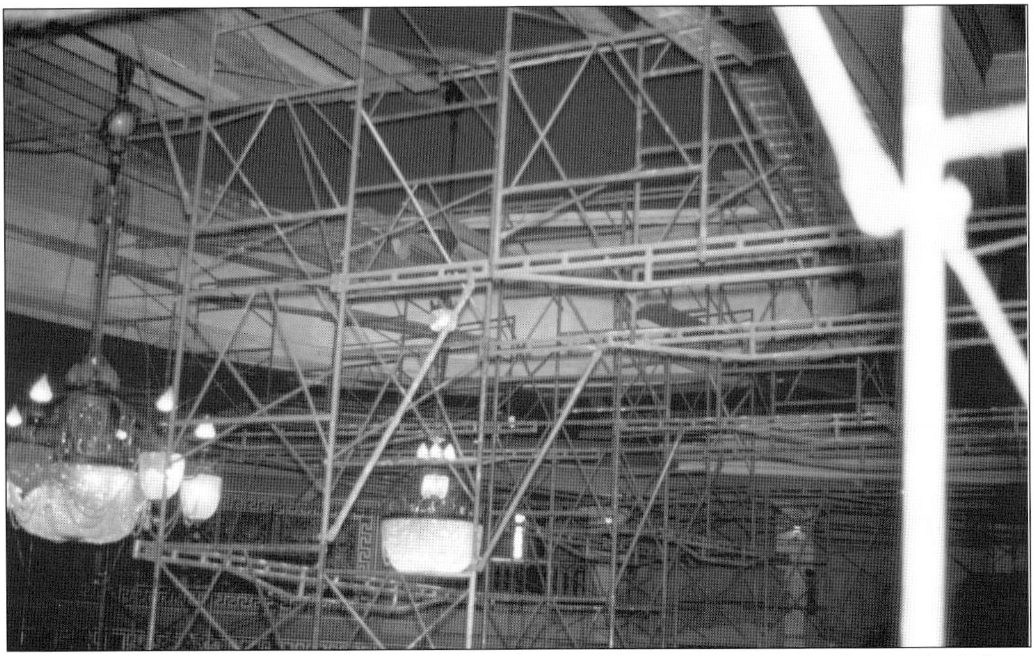

Repainting the Auditorium Ceiling. The auditorium demands constant upkeep to look fresh and current. The metal skeleton scaffolding rises more than 50 feet above the floor to repair the plaster and allow the painters adequate space to ply their trade. Contractors use extreme care to avoid damaging the three chandeliers in the ceiling.

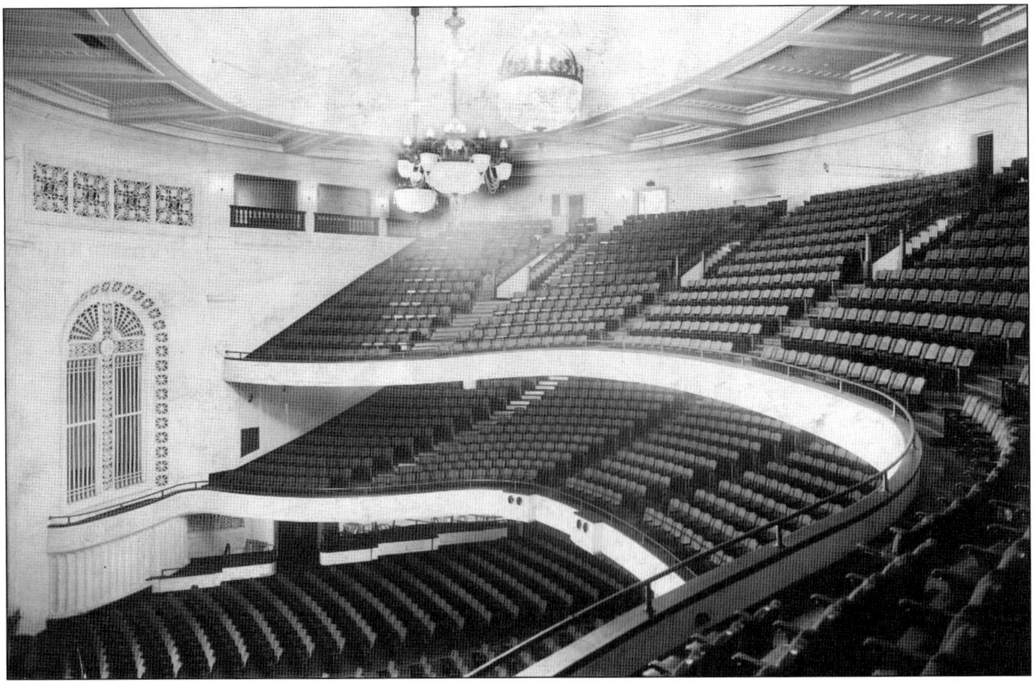

Another View of Auditorium. This almost 2,800-seat marvel remains one of Pennsylvania's finest entertainment venues. Wonderful views and fine acoustics make the facility a pleasure for both performers and audiences alike.

Six

Stage and Pipe Organ

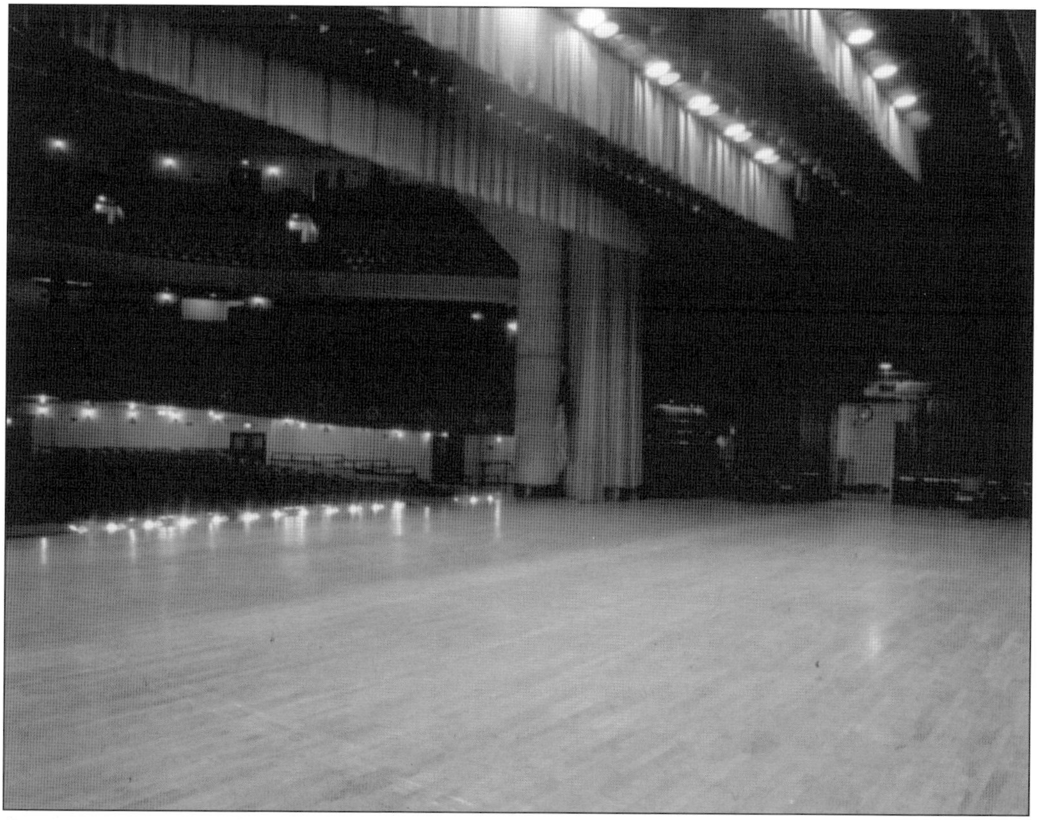

Cathedral Stage. The completed stage measured 80 feet wide by 40 feet deep with a height of 60 feet, and when new became the second largest between Chicago and New York. A total of 88 footlights, 165 border lights, and 40 stage lights illuminate the stage. The original rigging cost $11,300 in 1927, which would translate to approximately $150,000 today. The replacement value for the eight-foot Baldwin grand piano approximates $80,000, and the Moller pipe organ would exceed $1.5 million, based on current appraisals.

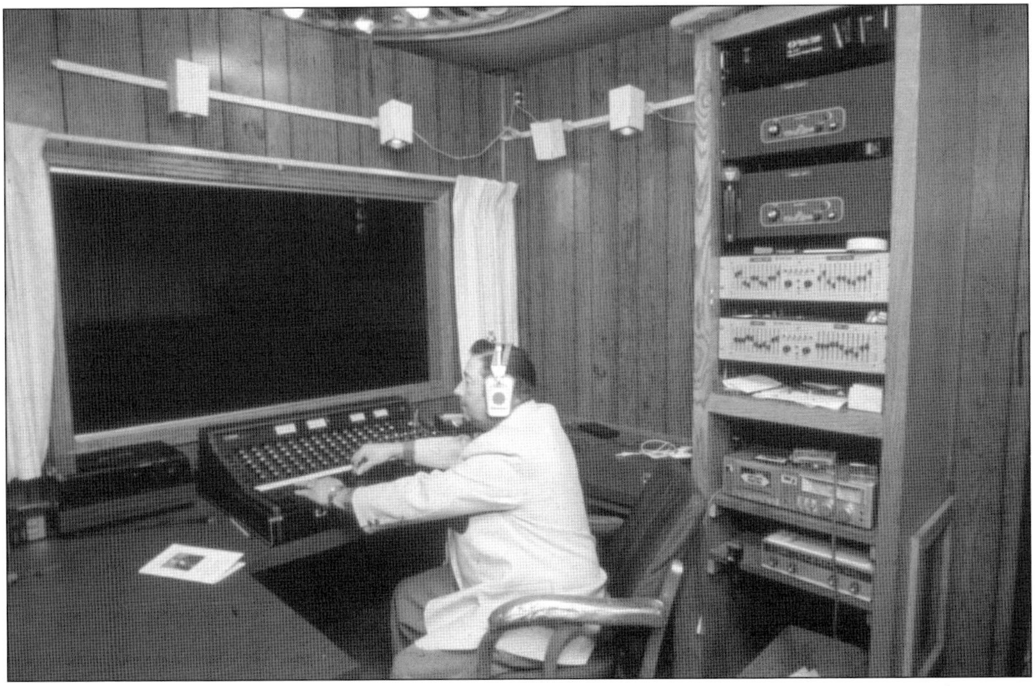

Sound Booth. Sound control technician Richard Heim looks over the auditorium to the stage as he adjusts the controls on the equipment. Although this photograph was taken in the 1980s, much of the equipment is still in use today and requires a high level of skill to operate.

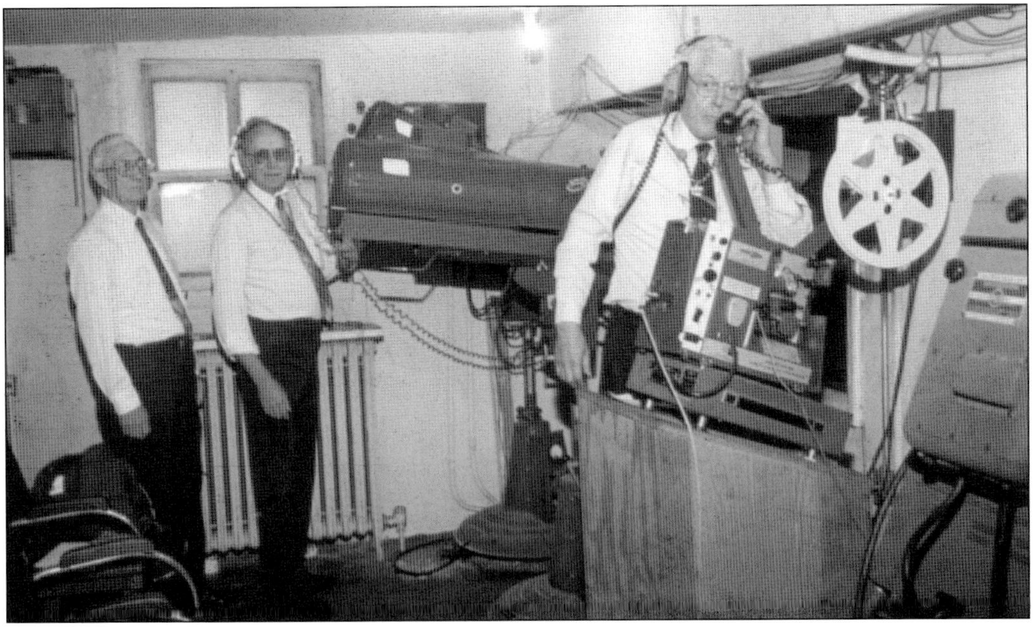

Projection Booth from the 1960s. Dale McConnell talks on the phone to the stage crew during a performance. His associates Larry McCarther and B. Edward Boyd line up a carbon-arc follow-spot while posing for a photograph. This equipment has been replaced over the years. On Saturday family nights, the auditorium filled to capacity for feature movies with cartoons, all shown on the projector pictured here.

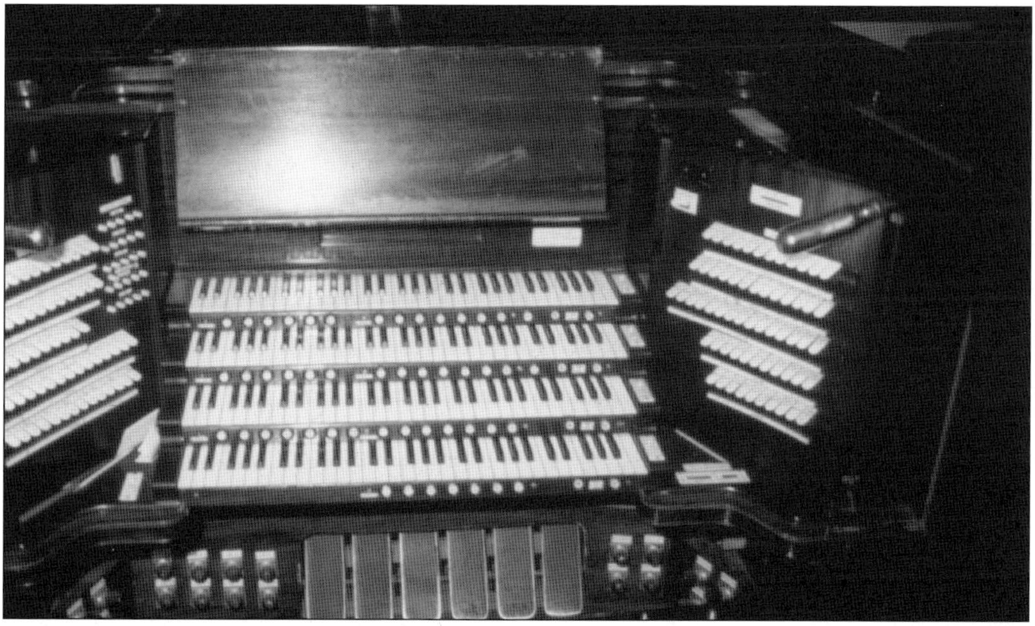

MAGNIFICENT MOLLER. On October 14, 1925, George Greer signed the agreement for a state-of-the-art $45,000 electro-pneumatic, four-manual pipe organ with M.P. Moller, Incorporated, of Hagerstown, Maryland. Five pages of the contract specified the organ's details, including genuine ivory and ebony keys. A recent Industrial Appraisal insurance replacement estimate valued the instrument currently at $1,542,600.

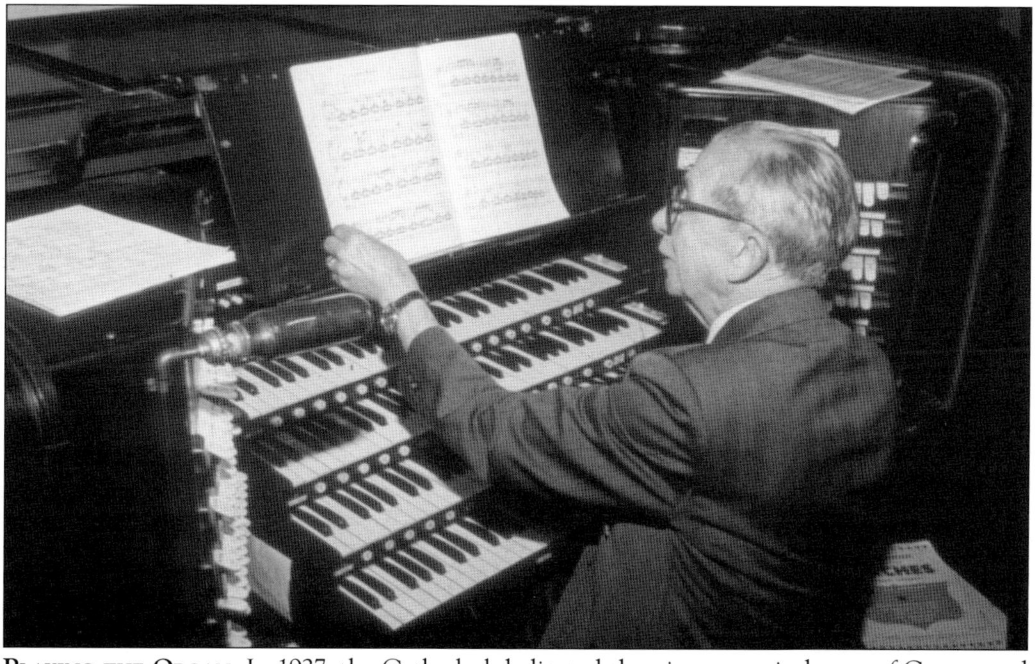

PLAYING THE ORGAN. In 1927, the Cathedral dedicated the pipe organ in honor of George and Charles Greer. A handsome bronze plaque at stage right commemorates the generous gift. The George Greer home at the corner of Jefferson and Lincoln Streets now houses the Lawrence County Society. Skilled organist Henry C. Knittel, 33rd degree, prepares to play for a performance.

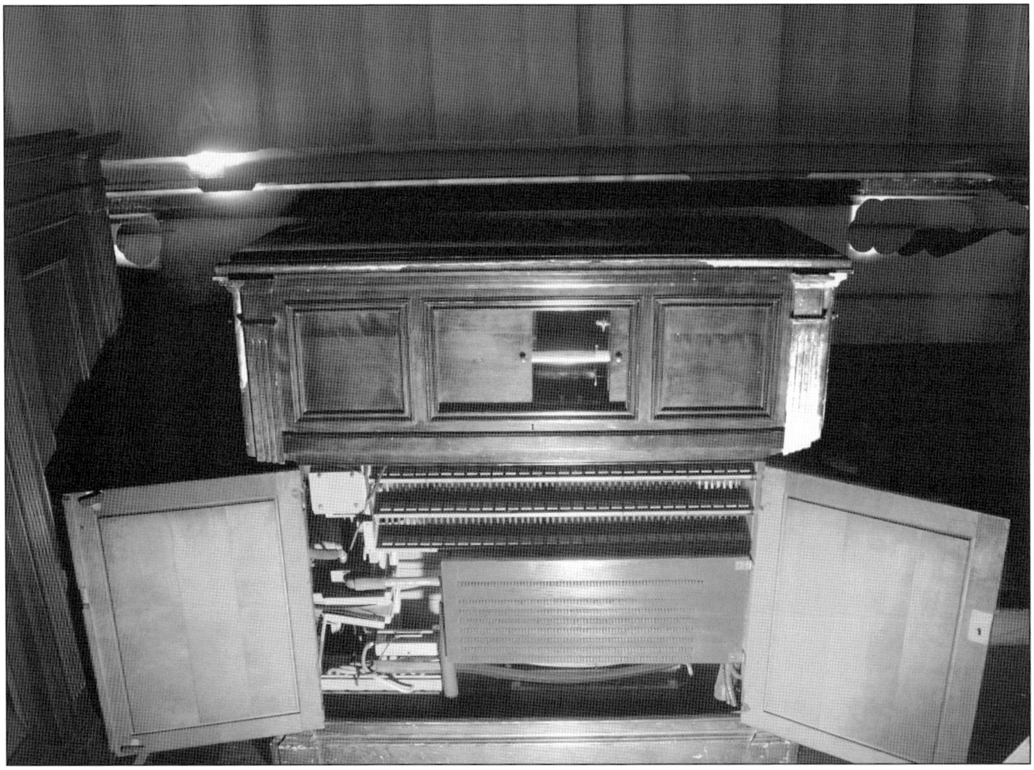

Solo Symphonist Player Organ. George Greer, an industrialist and cofounder of New Castle's tin mill, also donated a Moller automatic player and 12 rolls along with the organ at an additional price of $3,000, allowing for the use of instrument without a trained musician.

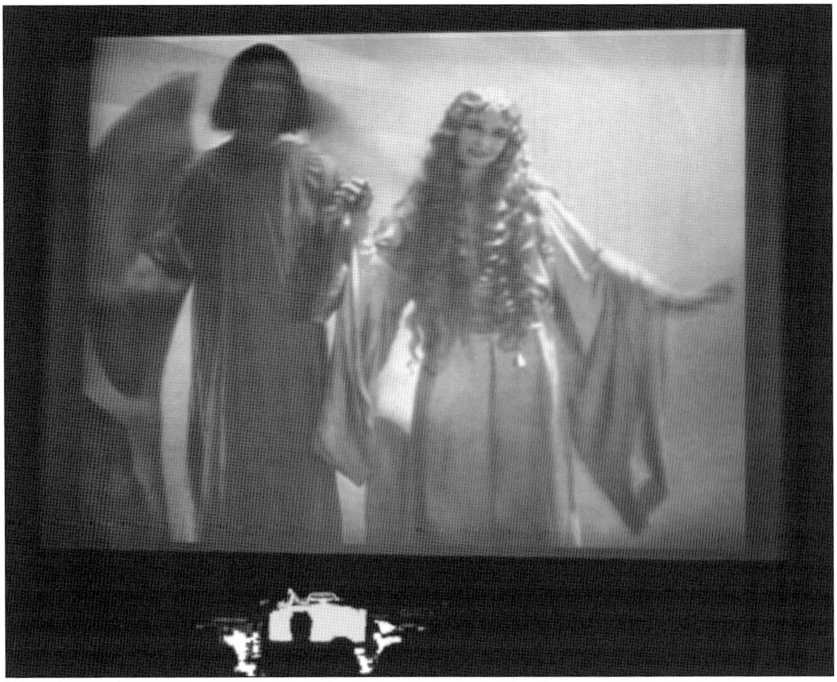

Phantom of the Opera. Christine Daaé performs in the famous silent, black-and-white movie based on Gaston Leroux's 1910 novel while world-renowned organist David Wickerham entertains the audience with the eerie sounds of the Cathedral pipe organ.

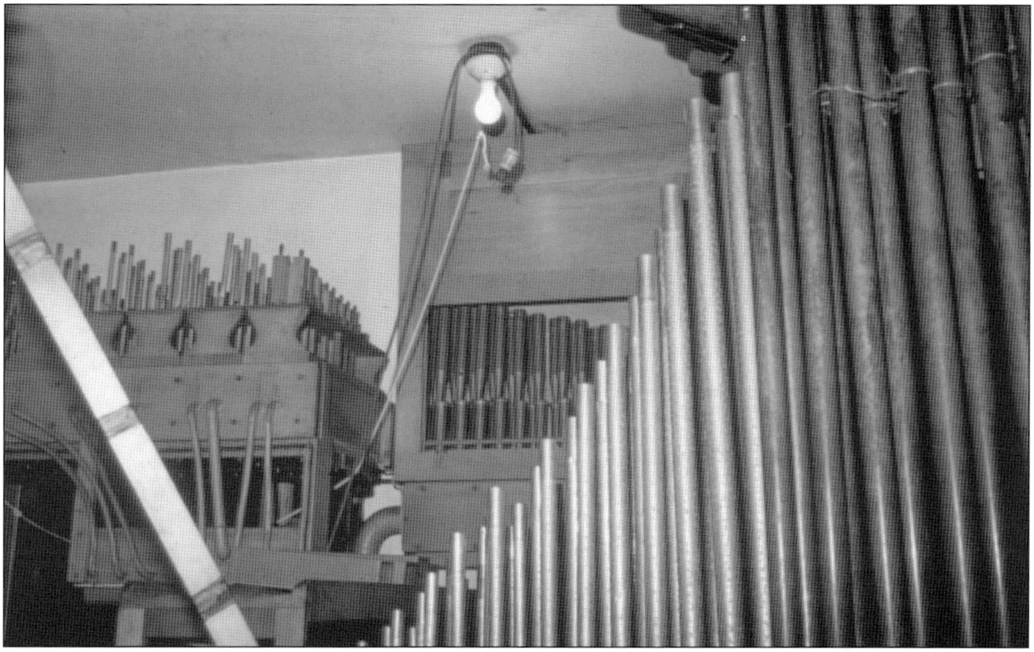

Organ Pipes Behind Decorative Grills. More than 3,700 metal and wooden pipes, measuring from mere inches to 32 feet in length, can be found on multiple levels and both sides of the auditorium, providing a phenomenal range of sound from the four-manual Moller organ.

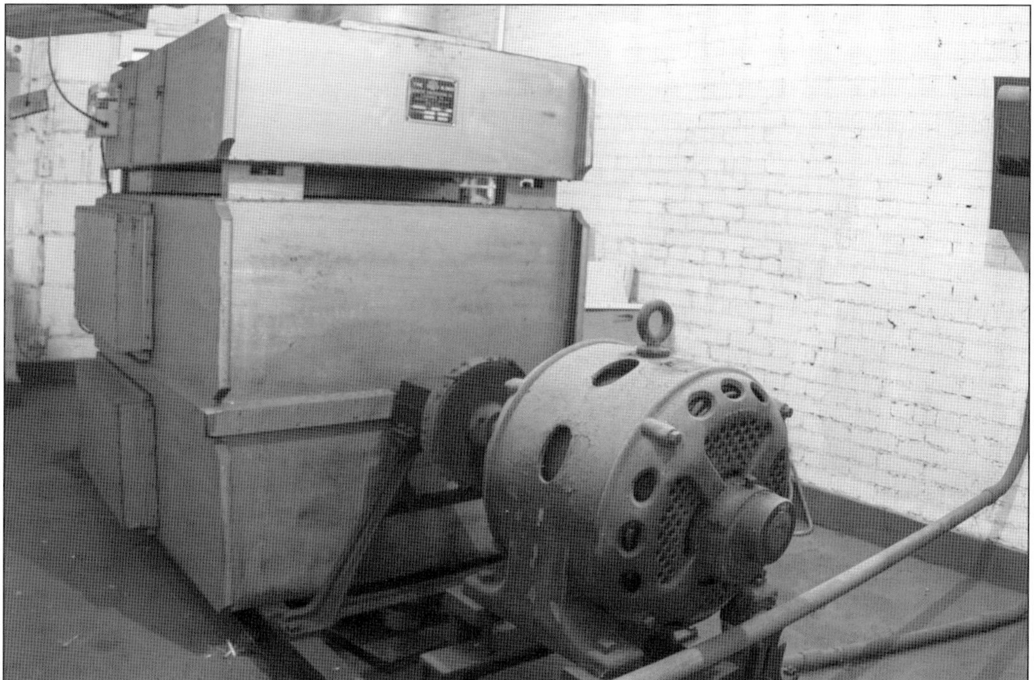

Blower Powering Organ. A Kinetic blower with a Century motor built in Lansdowne, Pennsylvania, under a 1915 patent provides 3,500 cubic feet of air per minute at 1,150 rpm. The unit splits the airflow to the pipes on either size of the auditorium.

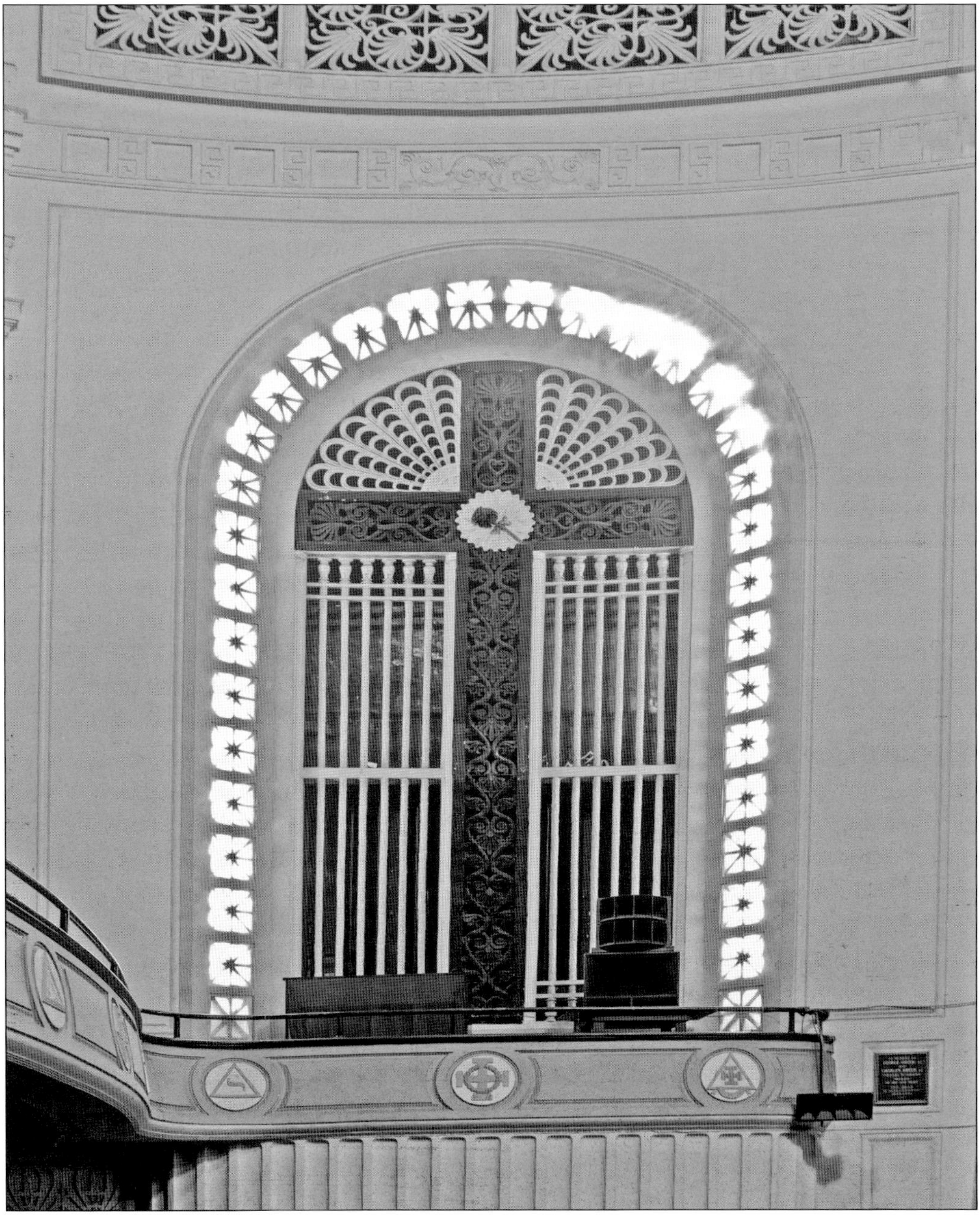

Organ Grill. A Rose Croix cross covers each of the two grills beside the pillars at the front of the stage facing out toward the auditorium. Louvers, naked to the viewer's eye, control the organ's volume. An upper grill above the arch hides the fixed volume pipes. A red, white, and blue lighted border adds drama for the audience. Giant speakers at the front of the grill radiate sound from the stage. A bronze plaque commemorates the Greer family gift of the organ and player to the Cathedral.

Philanthropist George Greer. This active Mason made ownership of the organ a reality. Brother Greer served as vice president of the First National Bank and co-organizer of New Castle Steel and Tin Plate Company, which later was merged into U.S. Steel in the early 1900s.

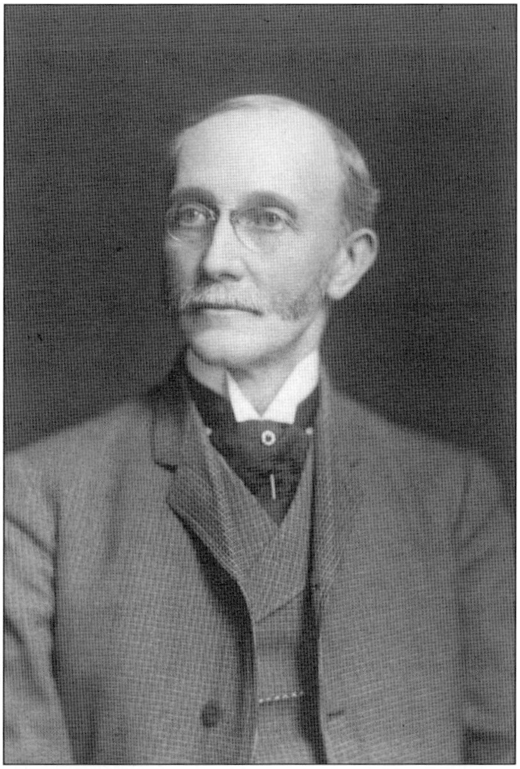

Twin Eight-Foot Grand Pianos. The Baldwin piano remains the oldest of the instruments and still is used occasionally. The building recently added a second eight-foot grand piano manufactured by William Knabe & Co. of Baltimore. The Cathedral owns seven smaller organs and six pianos as well as the Moller.

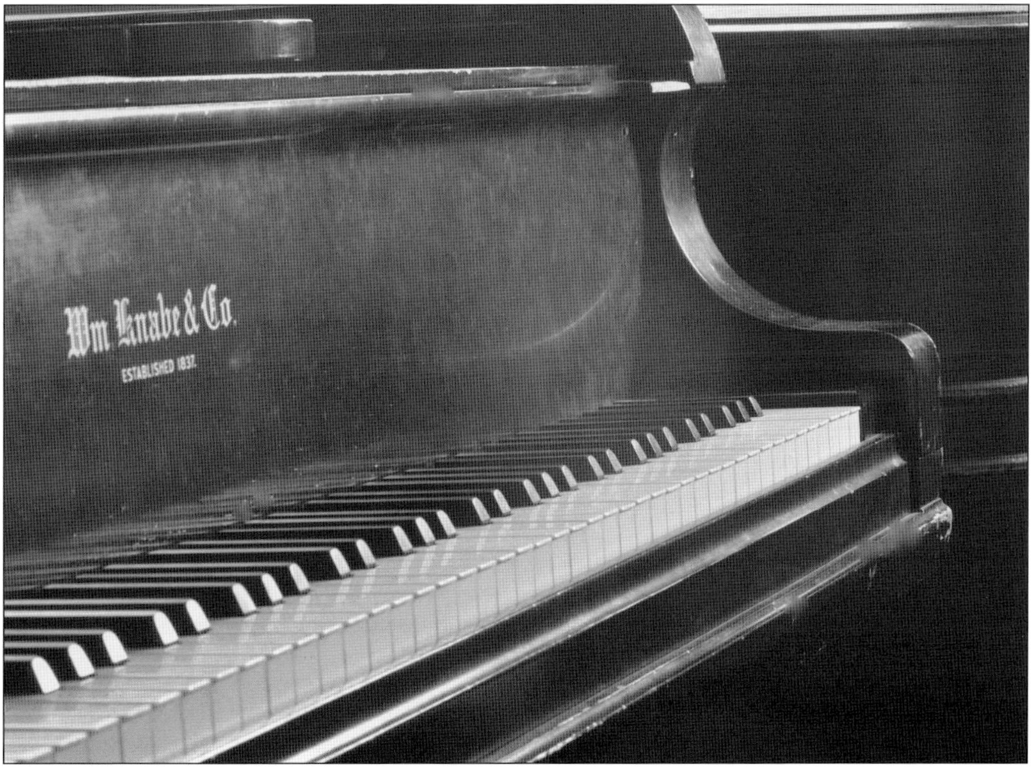

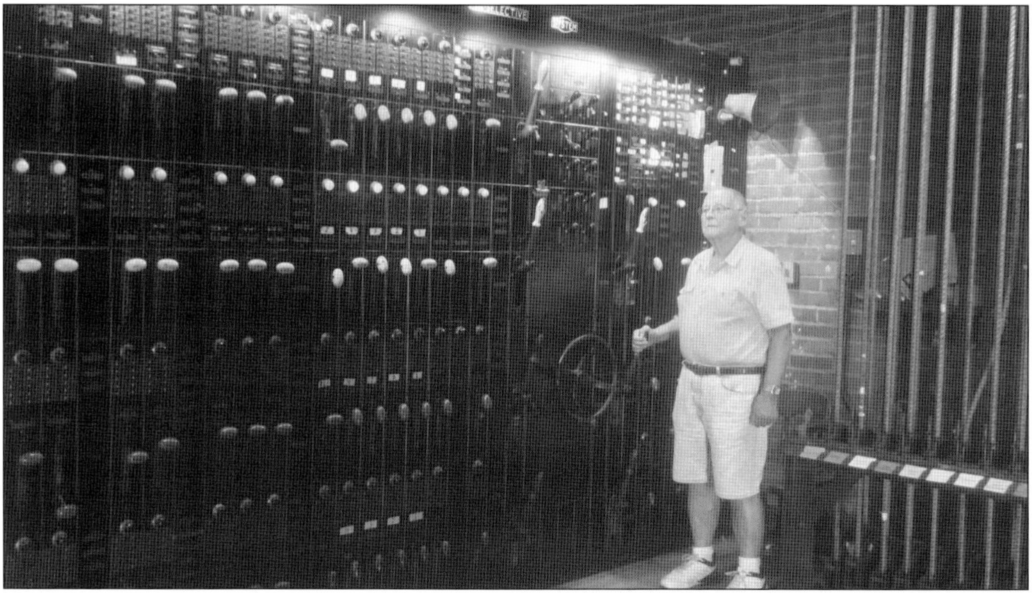

LIGHT BOARD. The stage still uses its original Frank Adam Company of St. Louis lighting system consisting of dozens of levers and nearly 2,000 switches. A major pre-selective function provides the ability to create 10 separate, multicolored lighting scenarios on the scores of border, foot, balcony-rail, and orchestra-pit lights. A variety of knobs on this 12-by-8-foot board provide the individual capability to dim illumination separately. For more than 45 years, stage manager Bob McKibbon has worked the lights.

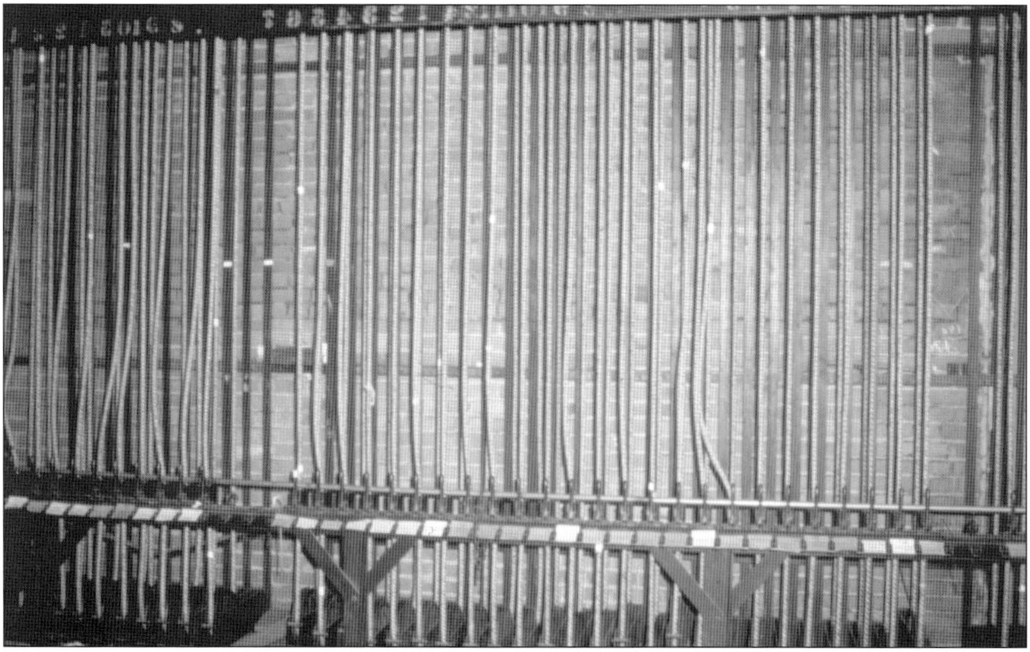

CURTAIN AND BACKDROP ROPES. Sixty ropes allow a broad array of backdrops, curtains, scenery, American flags, and several motion picture screens for a variety of uses in thousands of shows that have entertained hundreds of thousands of viewers over the decades on the 85-by-42-foot-deep stage—still one of the largest in the state.

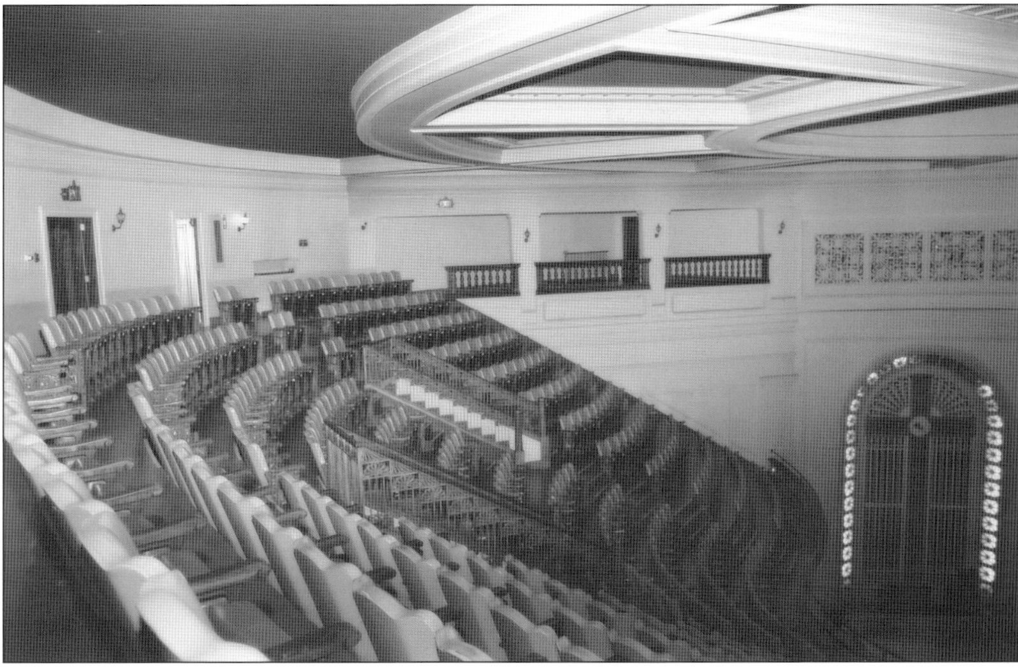

Another View of Auditorium. From almost every seat, the room sparkles with the light from the chandelier crystals, the handsome designs surrounding the walls and ceiling, the sconces along the rear, and the massive stage.

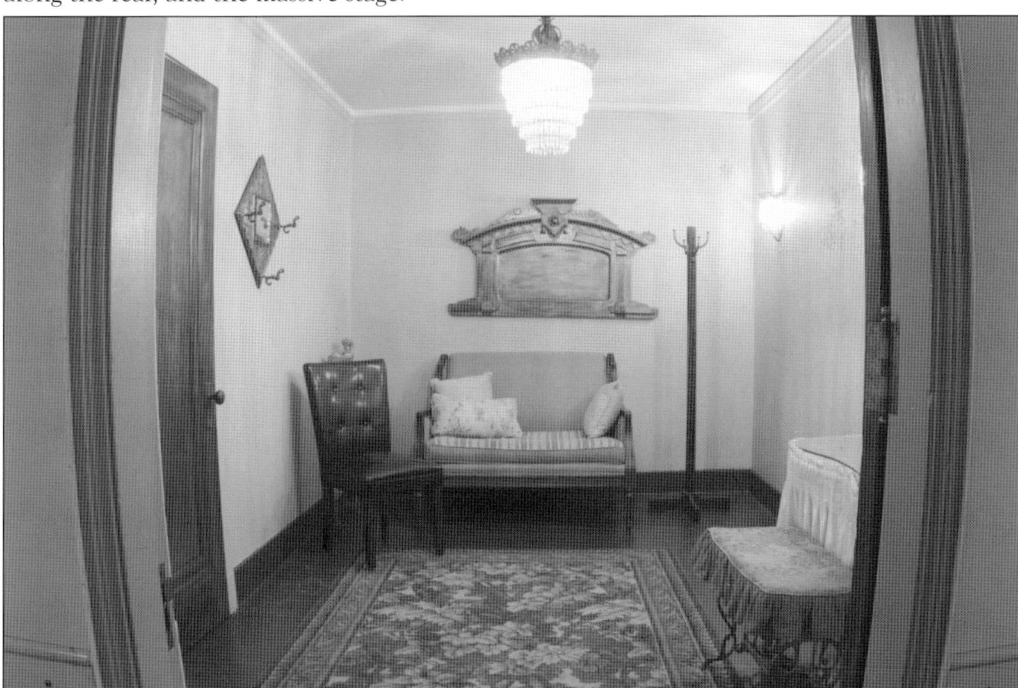

Star's Dressing Room. Imagine Reba McEntire, Johnny Cash, Kenny Rogers, Jerry Seinfeld, and local favorite B.E. Taylor preparing for another big show. Dozens of headline performers have used this dressing table prior to stepping on stage.

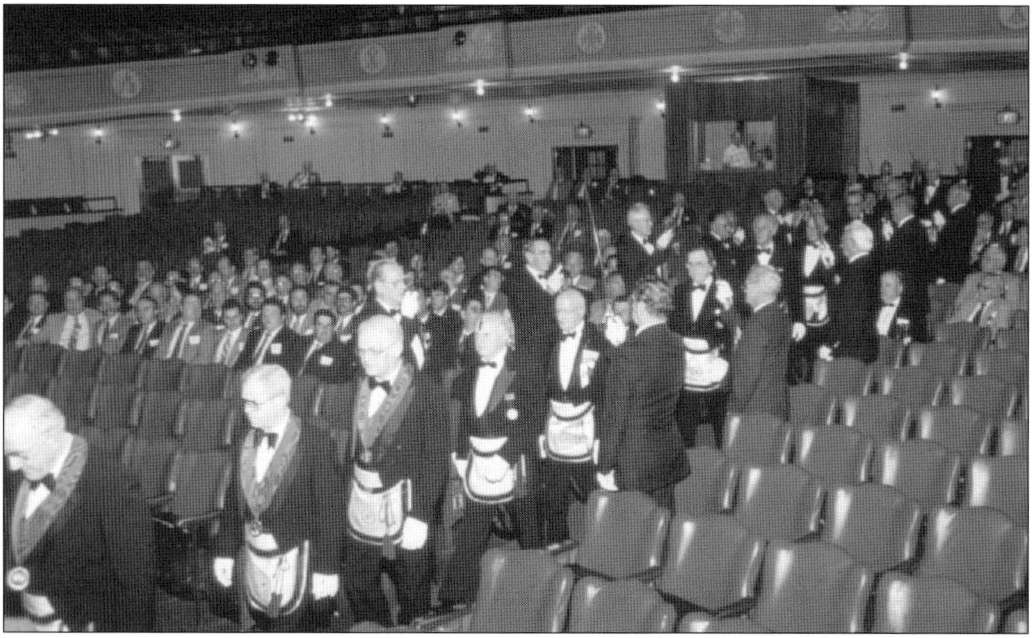

GRAND LODGE PROCESSION. District Deputy Grand Masters march below the crossed swords of the honor guards toward the stage while reunion candidates watch. Special collars and aprons identify their level of station.

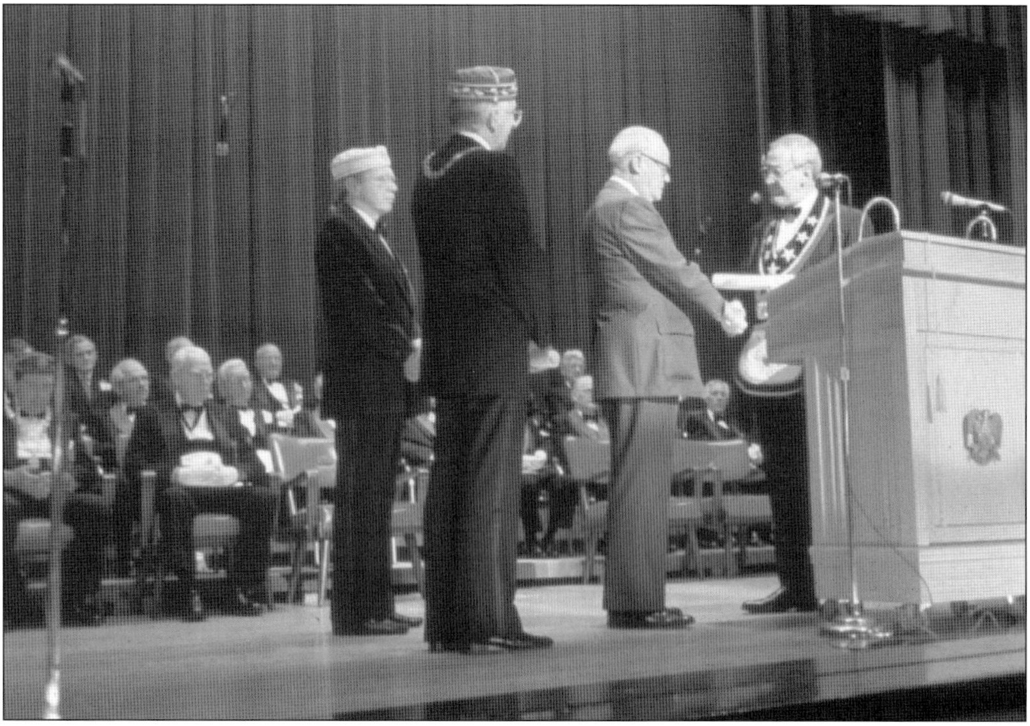

GRAND LODGE OFFICER MAKING PRESENTATION. Drew W. Washabau, Active and Deputy for the state of Pennsylvania, and Consistory Commander in Chief William L. McCarrier watch a congratulatory handshake during a reunion held sometime around 1989.

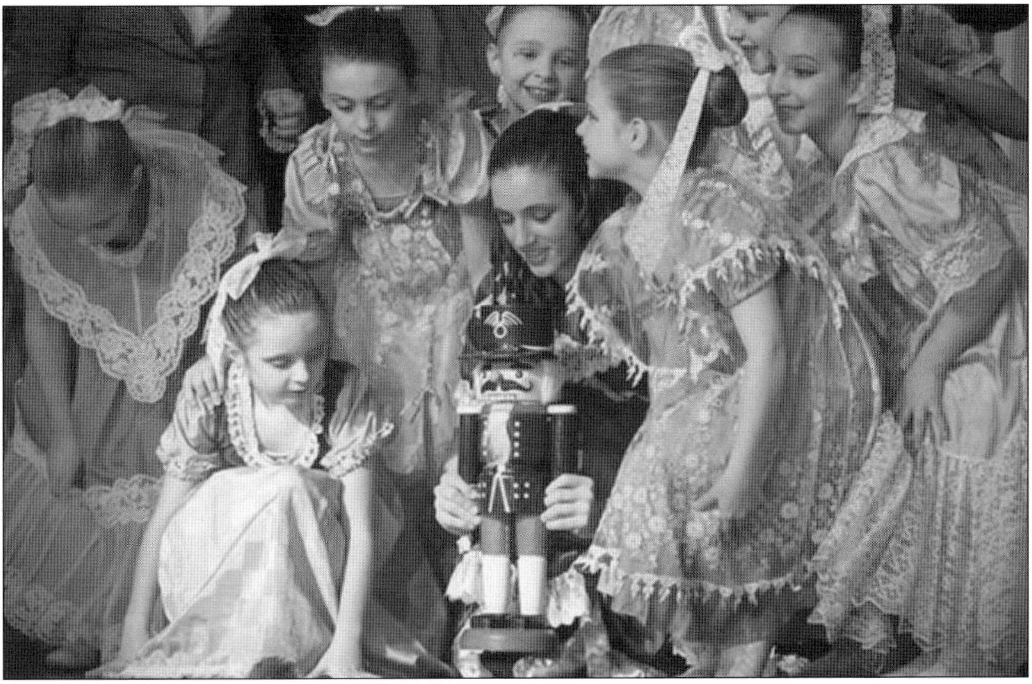

New Castle Regional Ballet. The New Castle Regional Ballet presented its first production of *The Nutcracker* at the Cathedral in December 1987 under the direction of Debbie Parou and choreographer Mansur Kamaletdinov. A cast of more than 100 perform each year. Whether cast as a mouse or the Sugar Plum Fairy, each participant learns discipline, teamwork, and respect. During the Act I party scene, Clara looks admiringly at her beloved nutcracker.

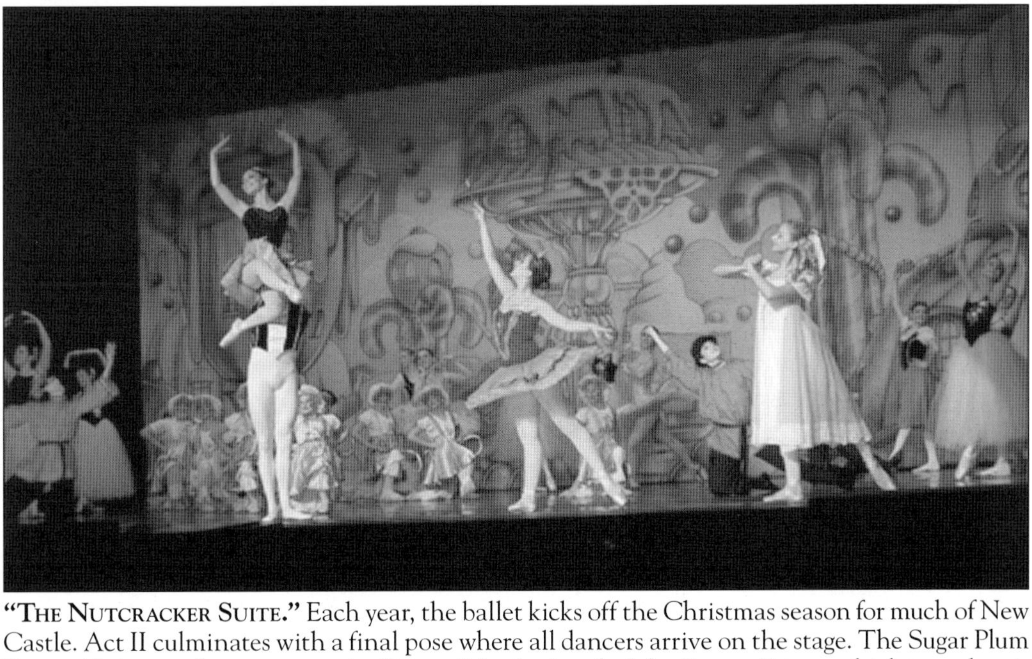

"The Nutcracker Suite." Each year, the ballet kicks off the Christmas season for much of New Castle. Act II culminates with a final pose where all dancers arrive on the stage. The Sugar Plum Fairy adds her enchantment as the Prince lifts the Land of the Sweets Princess high into the air while tiny Clara looks on adoringly.

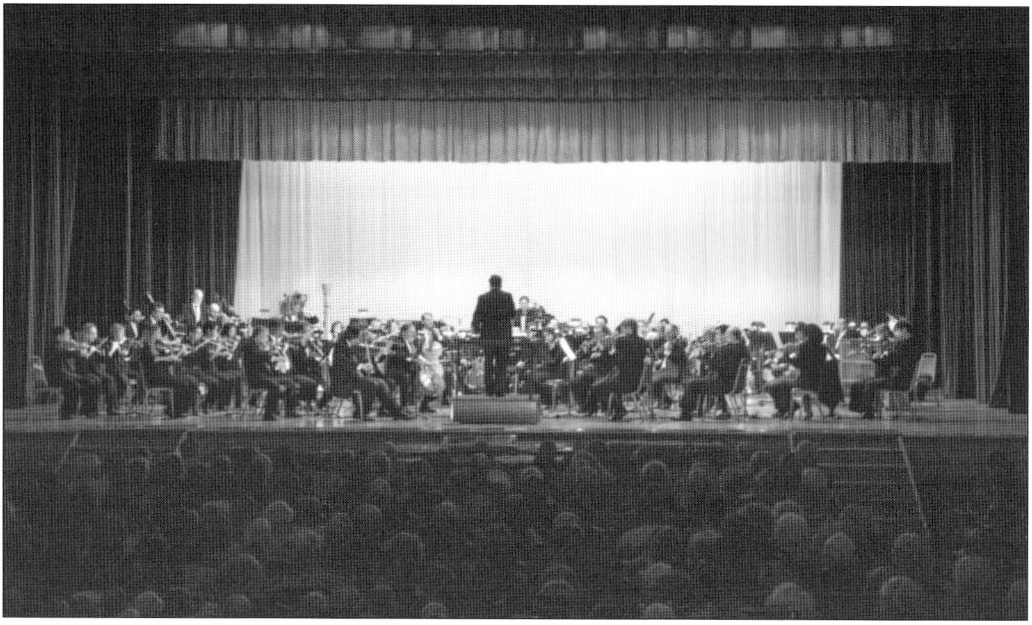

Pittsburgh Symphony Orchestra. For 17 years, the Pittsburgh Symphony Orchestra graced the stage of the Cathedral. Many remember Marvin Hamlisch, one of only two people to ever earn an Emmy, Grammy, Oscar, Tony, and Pulitzer Prize. Marvin served as the primary pops director from 1995 until his passing in 2012. Audiences loved his wonderful sense of humor, Coney Island hot-dog jokes, and winning personality.

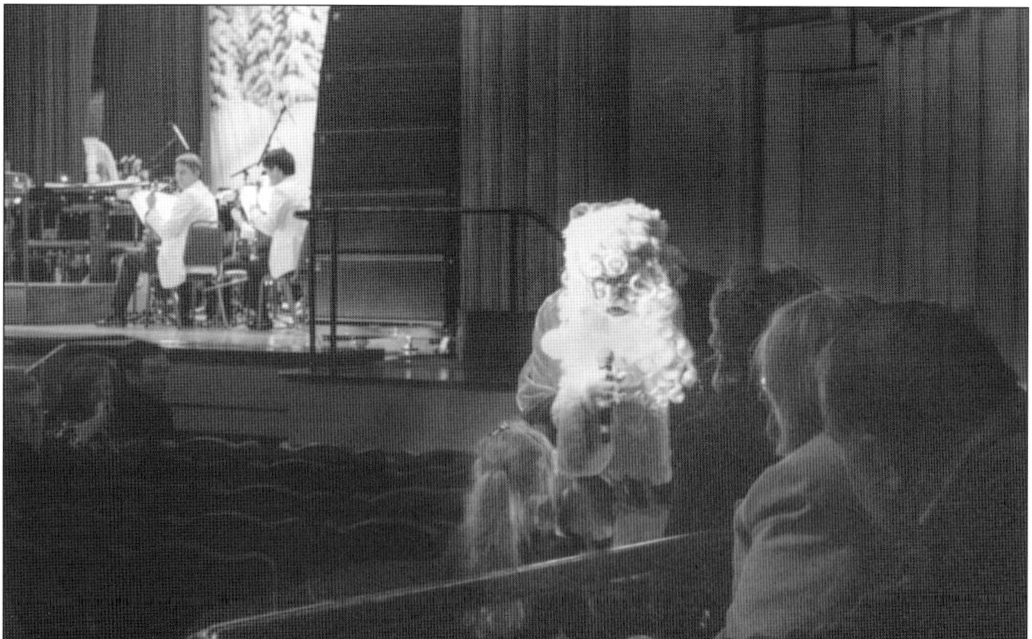

Pittsburgh Symphony Christmas Concert. A surprise guest entertains the audience at the 2016 holiday extravaganza. Santa delivers candy canes, music, and good cheer to young and old alike.

Seven

Kitchen and Dining Areas

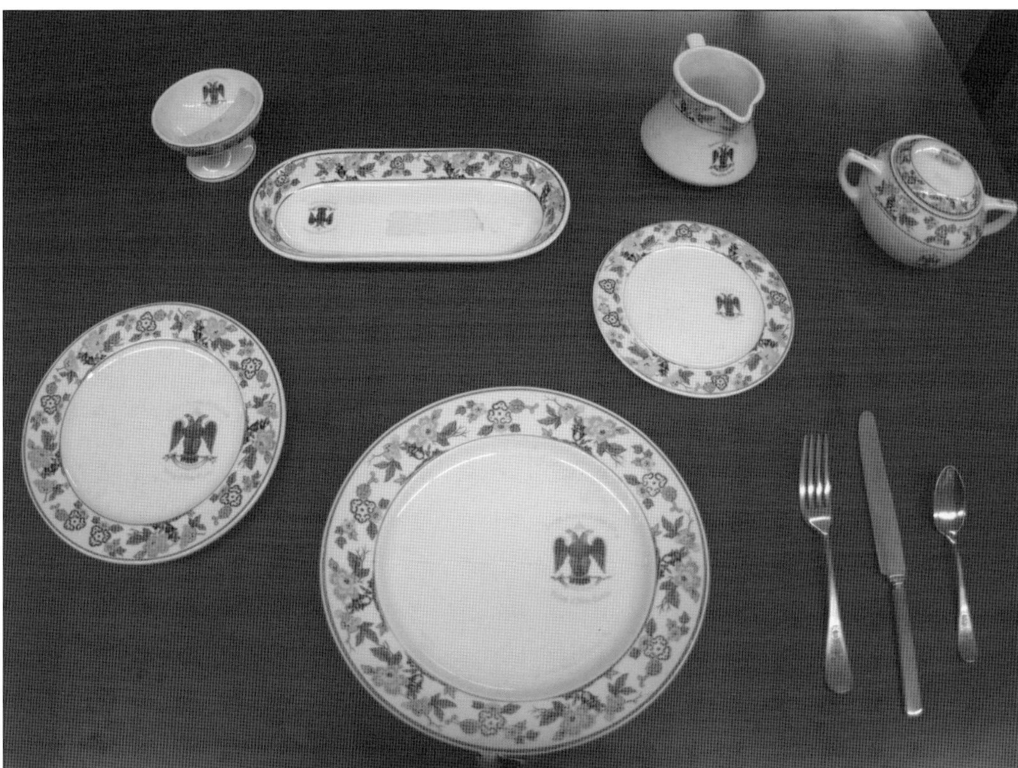

Shenango China Place Setting. The kitchen and dining facilities of the Cathedral represent an amazing array of equipment and serving pieces. Over the years, the Wimodausis ladies have served tens of thousands of meals to hundreds of businesses, wedding parties, special events, Masonic groups, and civic organizations. Kitchen specialist Julie Tomlinson recalls serving 1,500 guests in just 20 minutes. Such precision and timing proved the rule rather than the exception. The Shenango China Company produced all of the cups and dishes locally in New Castle.

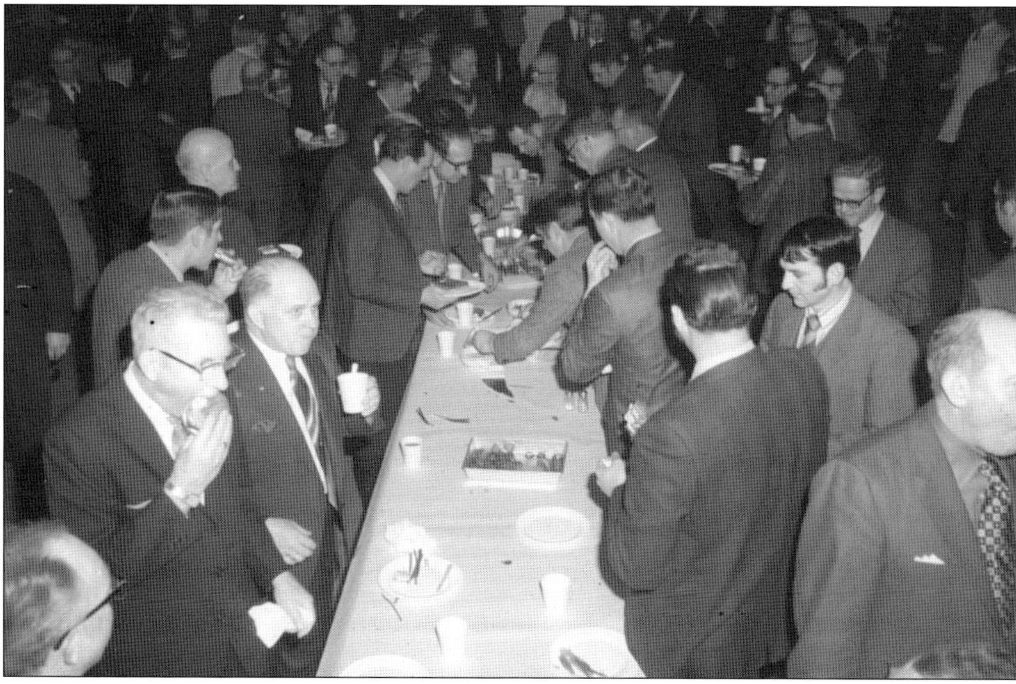

Fish Fry Event. Hundreds of members stood for the annual feast in the 1970s. All attendees dressed in suits and ties. Eventually, the older members and guests asked for and received seating, and the formal dress disappeared.

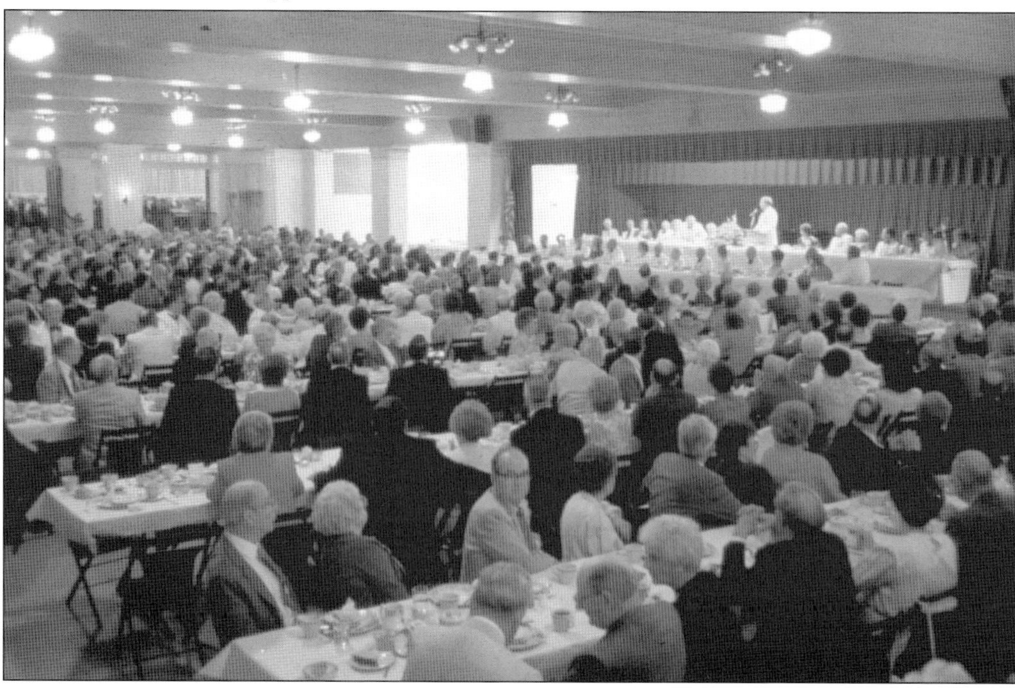

Consistory Reunion Banquet. The Wimodausis served with precision nearly 1,000 recently initiated new members, family, friends, cast, and officers. An overflow of attendees also fills the east and west dining rooms.

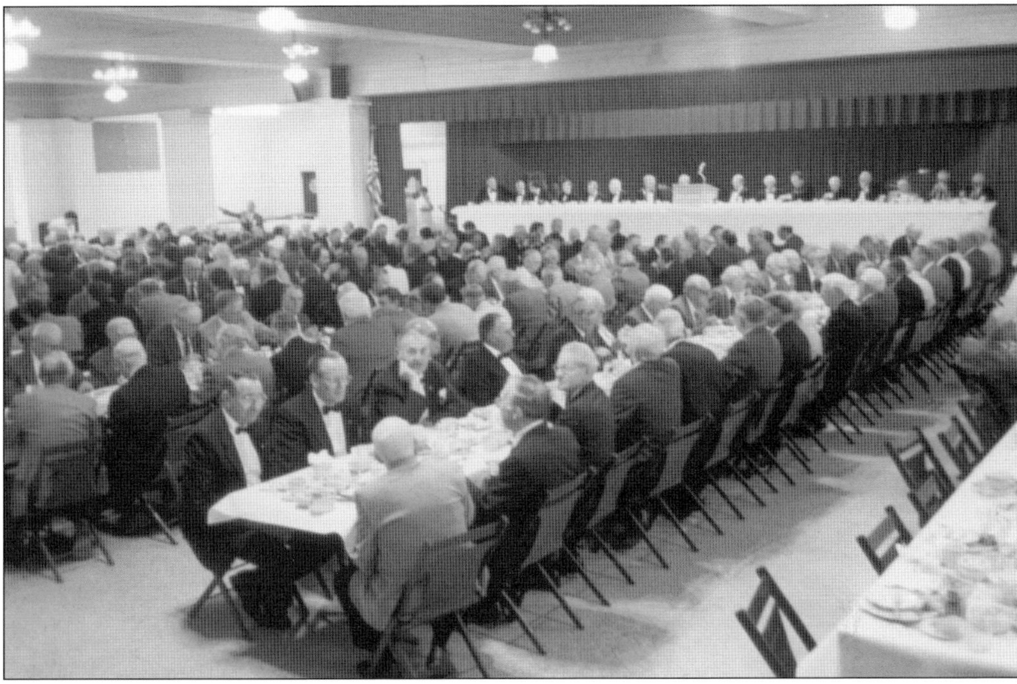

Men's Reunion Dinner. Wimodausis served hundreds of wonderful freshly cooked dinners on Shenango China plates and white tablecloths concluding a successful Valley of New Castle reunion in the main dining room. Dignitaries sat on the stage facing the audience.

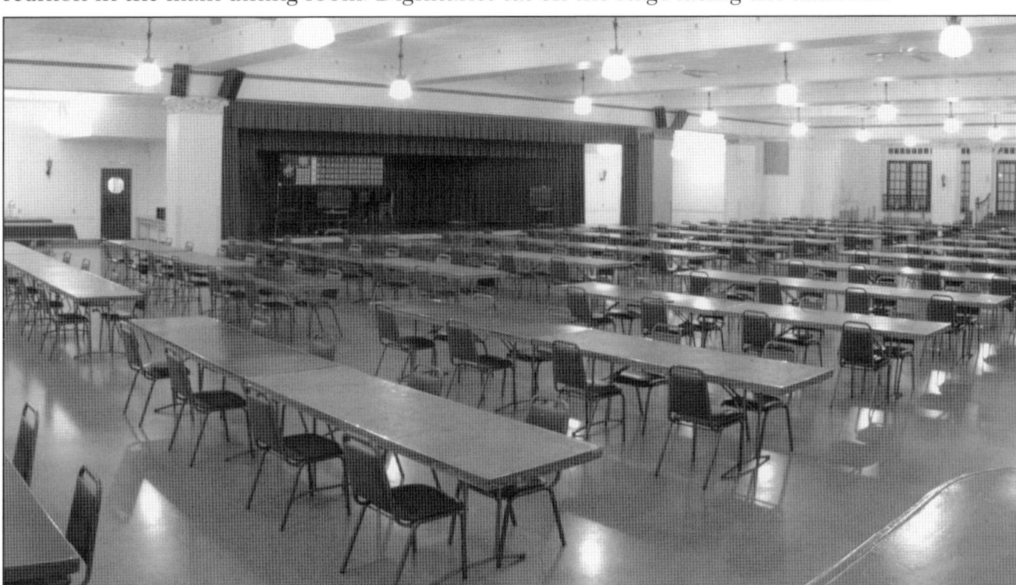

Main Dining Room. The center dining room, measuring 147 feet long by 58 feet wide excluding the stage and side serving zone, can seat 600 people comfortably for a dinner. Pink, grey, gold and oxblood interwoven-floral-and-phoenix-capped pillars surround the dining area. Terrazzo marble covers the floor. This 8,556- square-foot area, containing decorative sconces, six-sided chandeliers, and oyster half shell, side lights, has served a variety of functions. At the time this image was captured, the space had been set up for bingo.

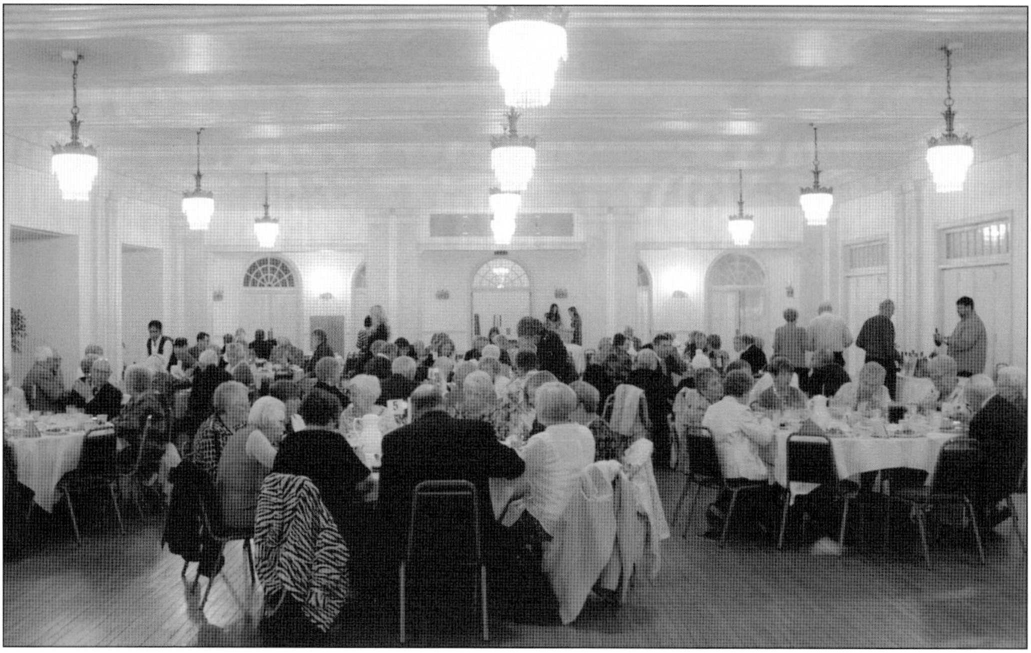

Crystal Ballroom. Light blue-grey and white walls, a maple floor, crystal chandeliers, greenery, and a compact stage make this 49-by-102-foot room the ideal venue for weddings, receptions, banquets, and dancing for up to 250 people at long tables or 150 at round tables. By opening the sets of glass folding doors, the nearly 5,000-square-foot room can expand into the main dining room.

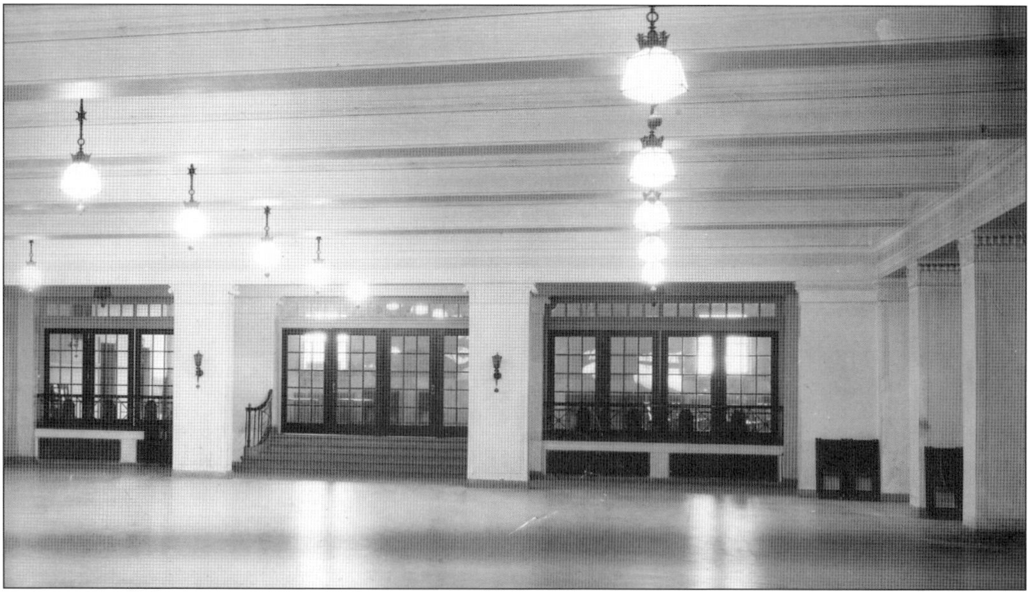

Looking into the East Dining Room. The 45-by-78-foot east dining room mirrors the west dining room in terms of size. The pillars match those in the main dining room. Ten milk-glass, domed chandeliers, and four sconces light the room. The 3,510-square-foot area presents an ideal location for smaller events and can be used for overflow dinner crowds of 250 people at long tables or 130 at round tables. Like the main dining area, the floor is terrazzo marble. These areas are also available for conferences, lectures, sports banquets, and craft fairs on a rental basis.

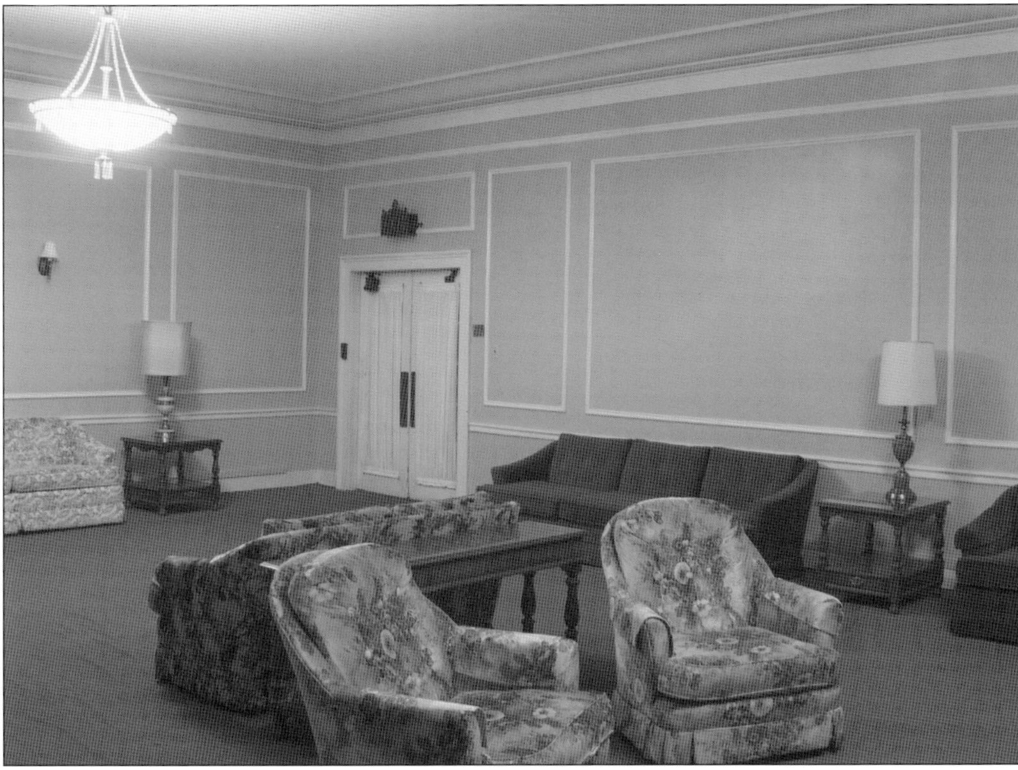

Imperial Lounge. Wimoduasis generously donated their time and funds to create a comfortable and attractive seating area for the ladies directly off the west dining room. This area presents an ideal rental location for smaller social gatherings such as card clubs, showers, and other primarily female-oriented functions. The facility adjoins a marble ladies' restroom. In addition, the Almira Foundation graciously donated the funds for separate family restrooms on each side of the Cathedral's dining level.

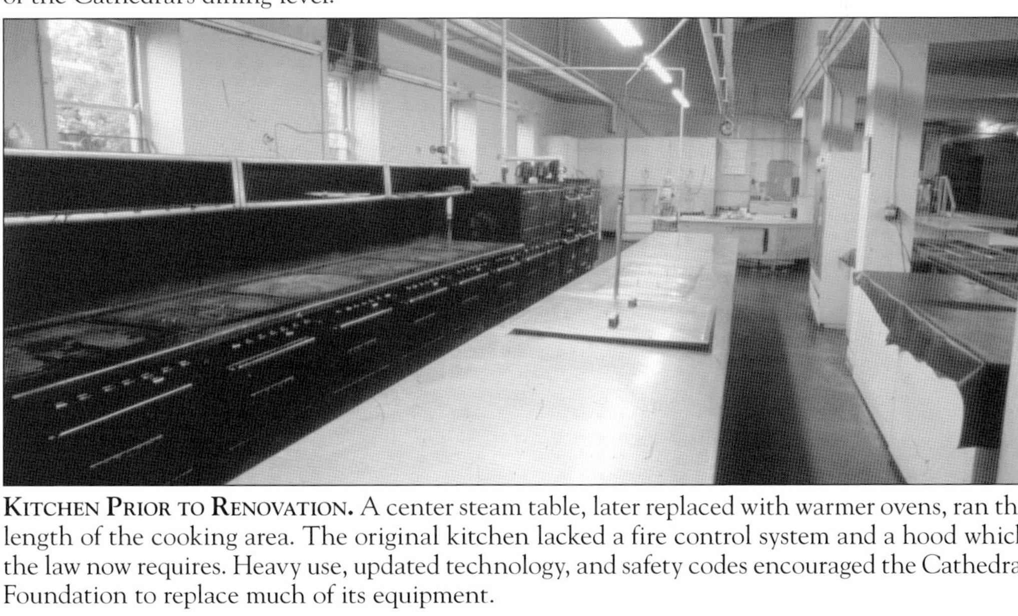

Kitchen Prior to Renovation. A center steam table, later replaced with warmer ovens, ran the length of the cooking area. The original kitchen lacked a fire control system and a hood which the law now requires. Heavy use, updated technology, and safety codes encouraged the Cathedral Foundation to replace much of its equipment.

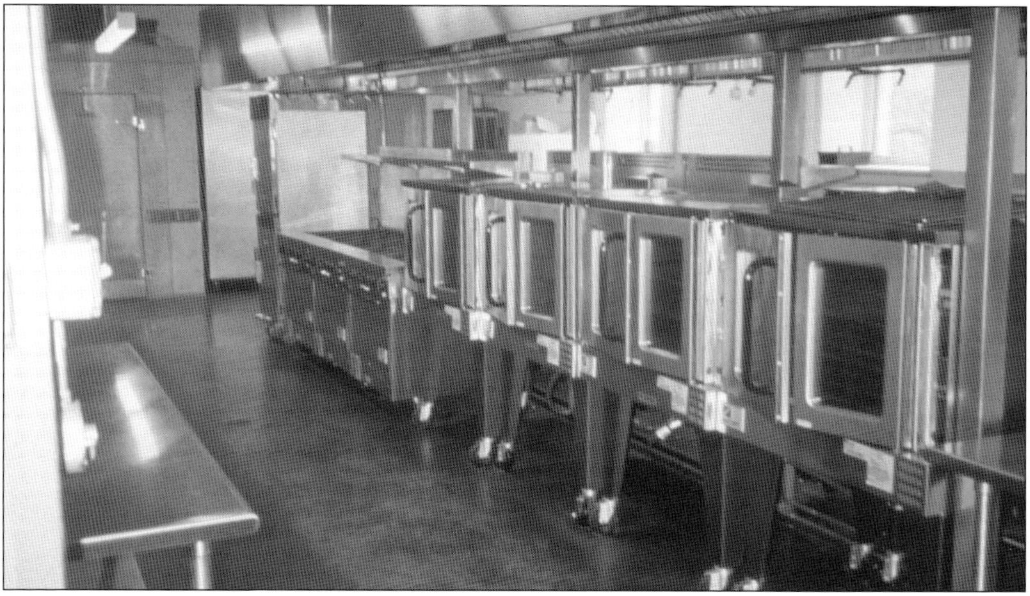

MODERN KITCHEN. The Industrial Appraisal Company of Pittsburgh estimated the replacement value of the kitchen equipment at nearly $250,000. This figure does not include the thousands of Shenango China dishes and serving pieces nor the array of silverware. Some of the kitchen highlights include a Univex stainless-steel potato peeler, eight Southbend stainless-steel, four-burner gas ranges; a stainless-steel, two-door food warmer; a Scotsman ice maker; a Hobart stainless-steel dishwasher system, and numerous refrigerators and freezers.

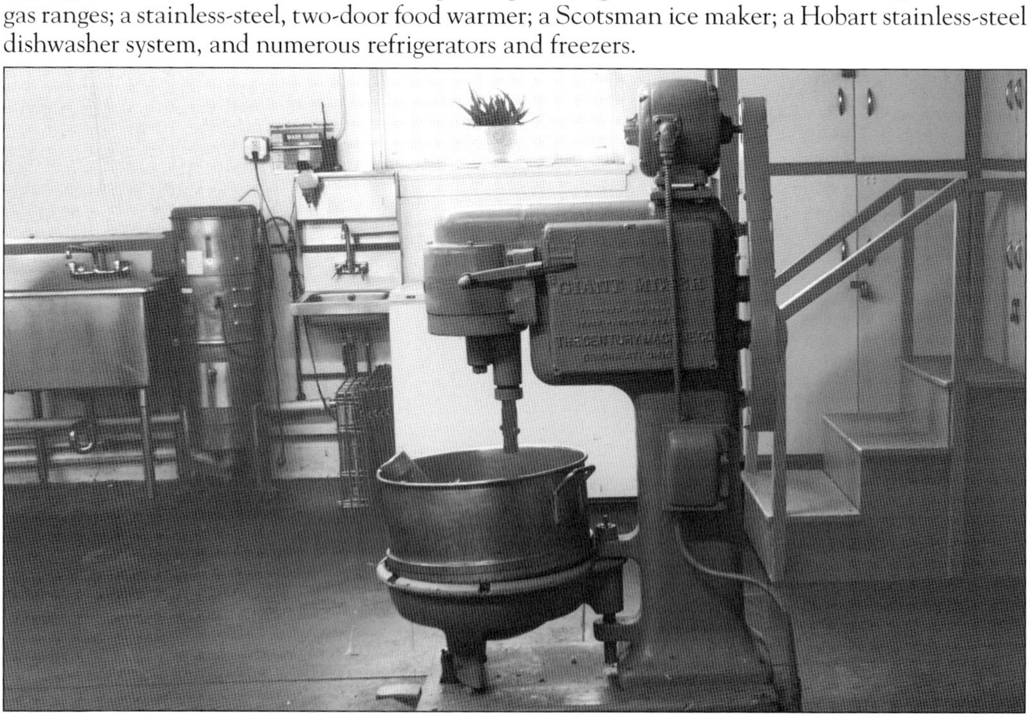

GIANT PASTRY MIXER. Wimodausis ladies made hundreds of pies using the Century Machine Company of Cincinnati's 50-quart capacity mixer based on a 1919 patent. Many old-timers remember Peggy Hill hefting huge sacks of flour, sugar, and other ingredients in just the right proportions to make tasty delicacies. Today, the mixer is used primarily for mashing potatoes.

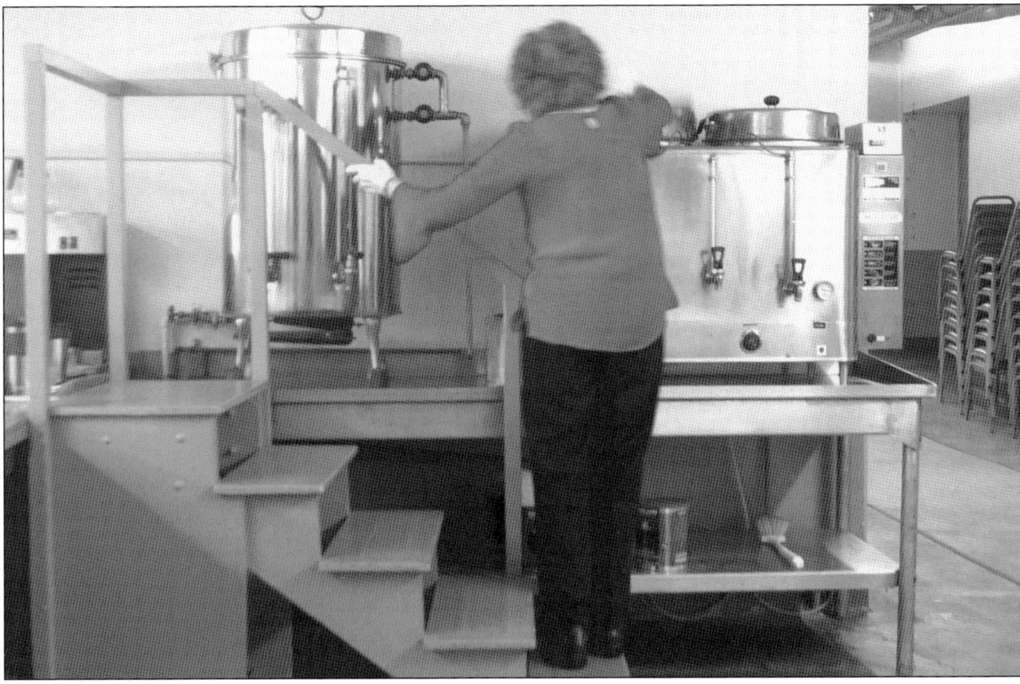

COFFEE URNS, GOOD TO THE LAST DROP. In addition to a commercial Bunn coffee maker and several smaller units, the kitchen sports a Cecilware 400-cup coffee urn and a second generic unit of equal capacity.

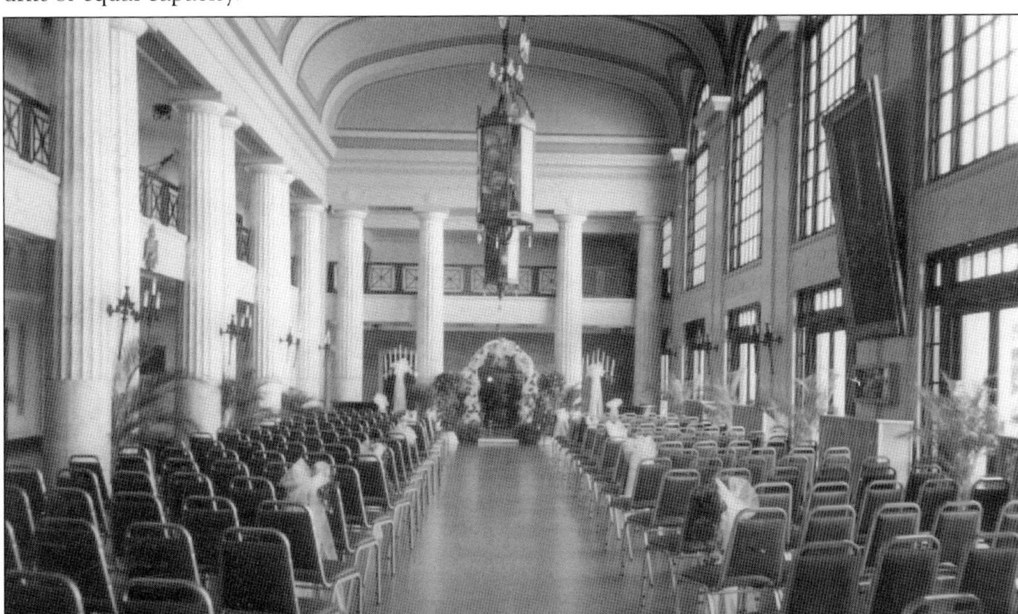

LOBBY TRANSFORMED INTO A WEDDING CHAPEL. Various rooms in the Cathedral offer the flexibility to handle dinner parties, graduations, confirmations, and a variety of dining uses on a rental basis. Cathedral event experts have converted the lobby to a wedding chapel complete with white lace trim, an altar, and greenery. This wedding accommodated more than 100 guests. For larger gatherings, the upper-level mezzanine provides space for additional seating.

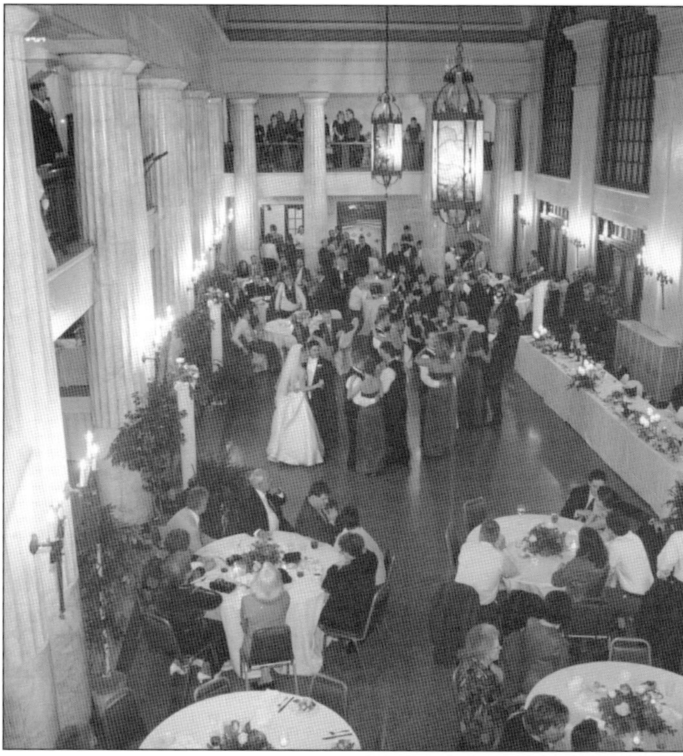

Lobby Wedding Reception. White table cloths, chair covers, napkins, and floral centerpieces transform the lobby into an idyllic location. Bathed in light from the massive windows and highlighted by the 18 pillars, this 2,800-square-foot room would make any bride and groom glow with excitement.

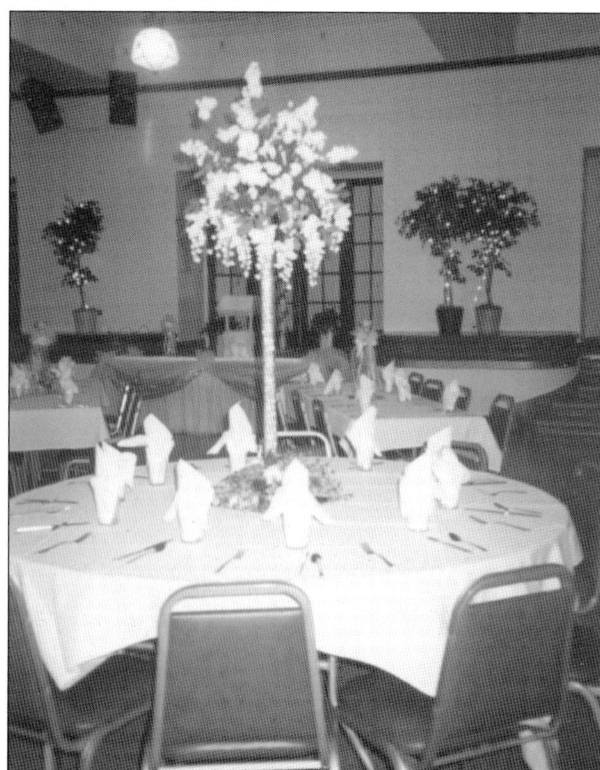

Decoratively Set Table. The main dining room awaits a wedding reception. Topiaries, fancy folded napkins, and freshly ironed white tablecloths add elegance to the room. A wishing well for gift envelopes sits in front of the glass doors.

Eight

Behind the Scenes and Curious Corners

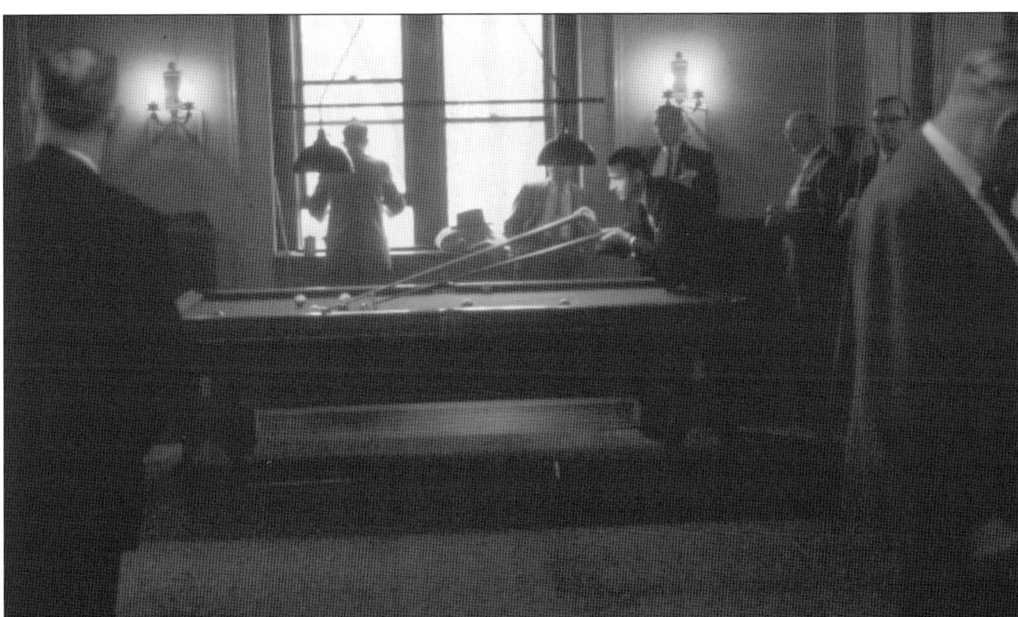

Special Degree Work. In 1960, the area, which later became the Dyslexia Learning Center of Western Pennsylvania, served as a recreational pool and card-playing center for the brothers. Candidates could relax between degree work and take full advantage of the facility following their initiation. The building's lowest level also contained a duck-pin bowling alley, later converted to a civil defense room, a shooting range for the Boy Scouts, and now, a dead storage area.

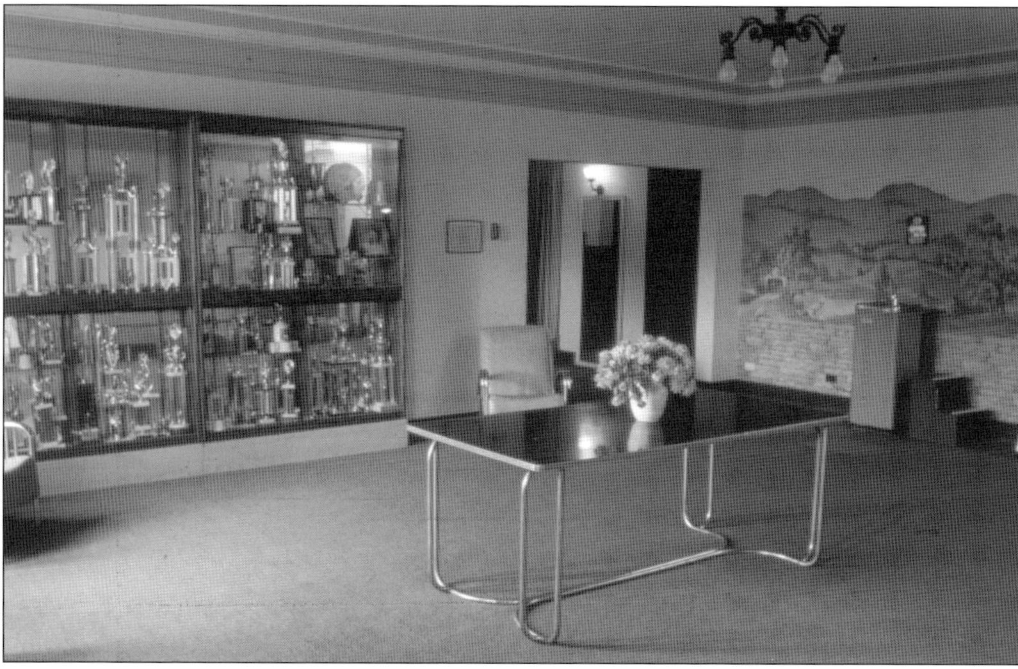

Lounge and Trophy Room. Adjoining the pool and recreation room, a lounge housed comfortable 1950s-style vinyl chairs, couches, and a water fountain. Competitive trophies filled an entire case abutting a muraled wall. Today, the Dyslexia Learning Center of Western Pennsylvania utilizes this space.

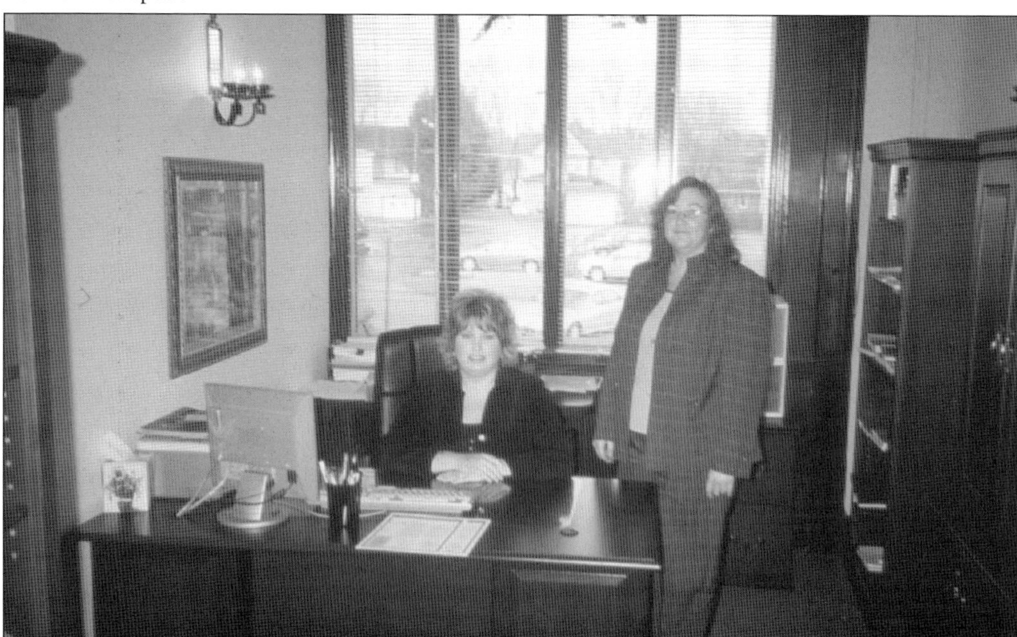

Learning Center. Director Melissa Merritt discusses educational strategy with one of her workers. The Dyslexia Learning Center of Western Pennsylvania provides free training for all attending students, employing the Orton-Gillingham Method of teaching to help children learn to read.

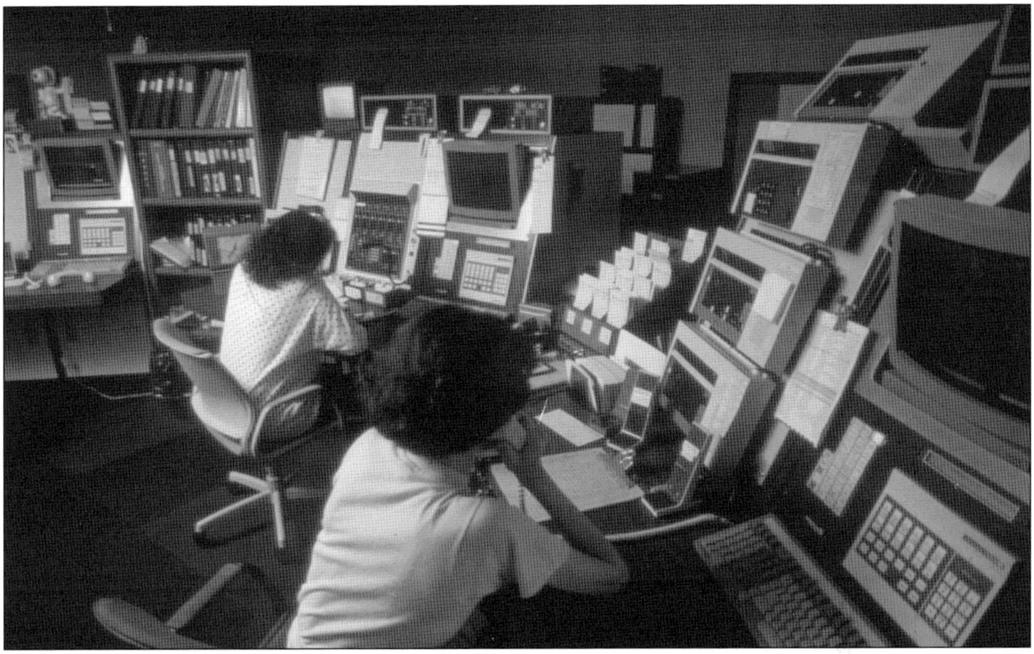

Lawrence County Emergency Center. For 34 years, the area's 911 headquarters sat on the top floor across the front of the Cathedral. Advanced equipment and skilled technicians coordinated with first responders and protected the community. Due to a need for additional space, the emergency center left for new headquarters in 2016. Prior to 911, the room served as the home for radio station WKST.

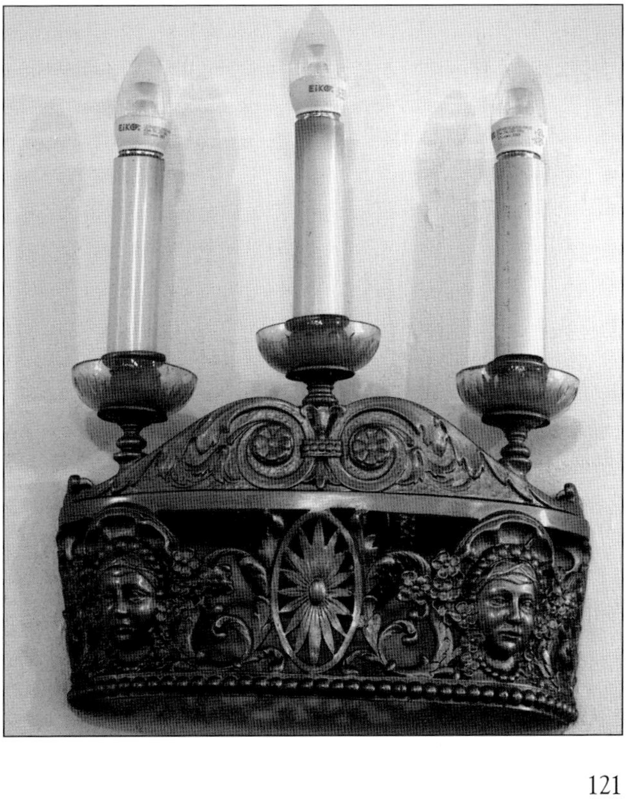

Dining Area Sconces. The Crystal Ballroom contains 12 golden sconces, each featuring a center rosette, two heads of regal ladies, and three electric candles. The food serving area beside the kitchen has four sconces which echo the design of the chandeliers located in the main dining room. Another four units brighten the steps leading to the east and west dining areas. The lighting photographed represents a small sample of the illumination found along the Cathedral walls.

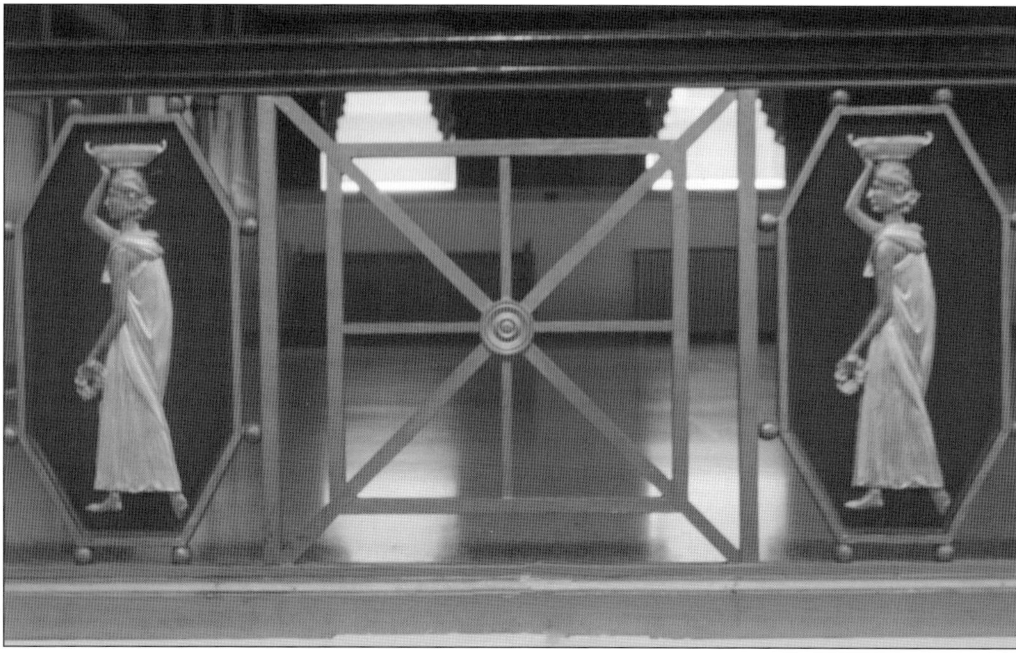

Decorative Railing in the Main Dining Room. Gold-colored Greek ladies in silver robes carrying bowls on their heads and wreaths in their hands separate the main dining room from the east and west dining areas. The railing design balances the classical feel and simplicity of the marble floors and beams.

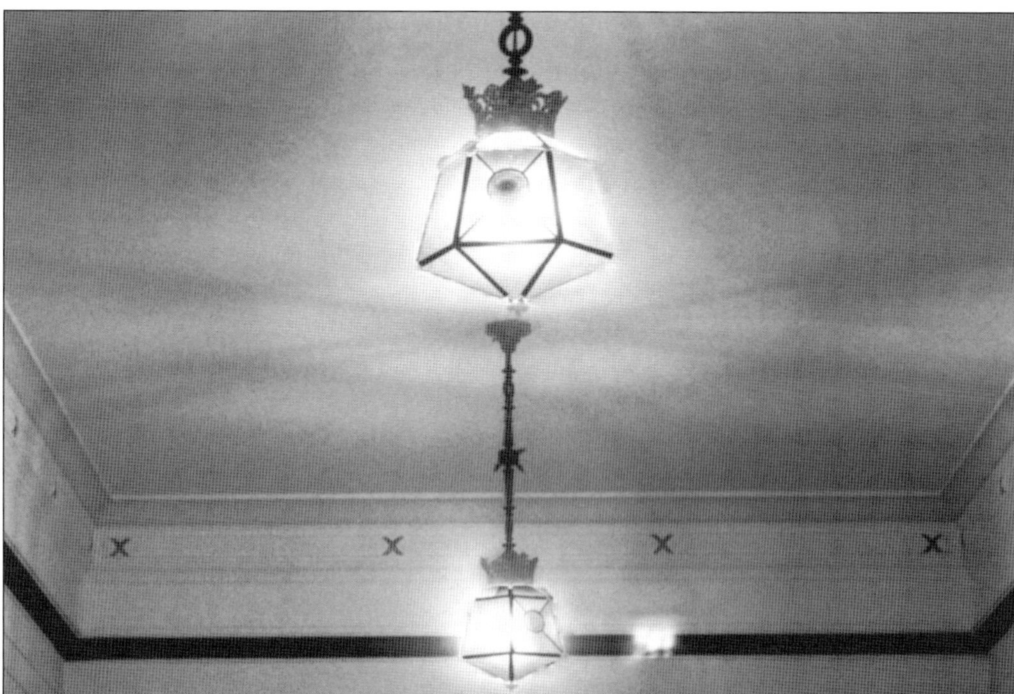

Main Dining Room Chandeliers. A total of 27 dome-shaped chandeliers incorporate frosted glass highlighted by blue, green, and yellow circles. Each of the fixtures hangs from a dark metal rod.

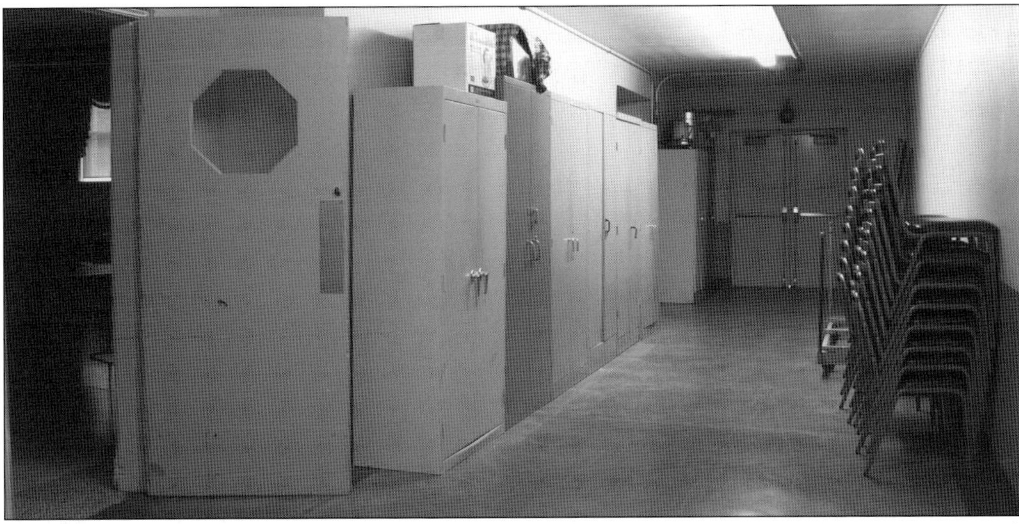

Unexplained Phenomena. Maintenance person Brady Robertson witnessed the appearance of a young boy in a blue coat running from the east dining room through a solid wall across the hall. Another sighting involved a specter of a young boy speeding across the lobby floor into an open elevator, his footprints plainly marked on a freshly mopped floor. A maintenance person warned, "Don't run. The floor is slippery." Upon close inspection, tiny footprints remained, but the open elevator stood empty. When Rob Cummings, while working on the stage, heard a high-pitched laugh coming from the rear of the first balcony, he spotted a boy-sized dark figure run across the back aisle of the dimly lit auditorium balcony. Cummings called out, "Please slow down. It's dark. You're going to fall." He immediately turned on the lights and went up the stairs to unlock the balcony area. All remaining doors in the balcony were locked, and the apparition had vanished.

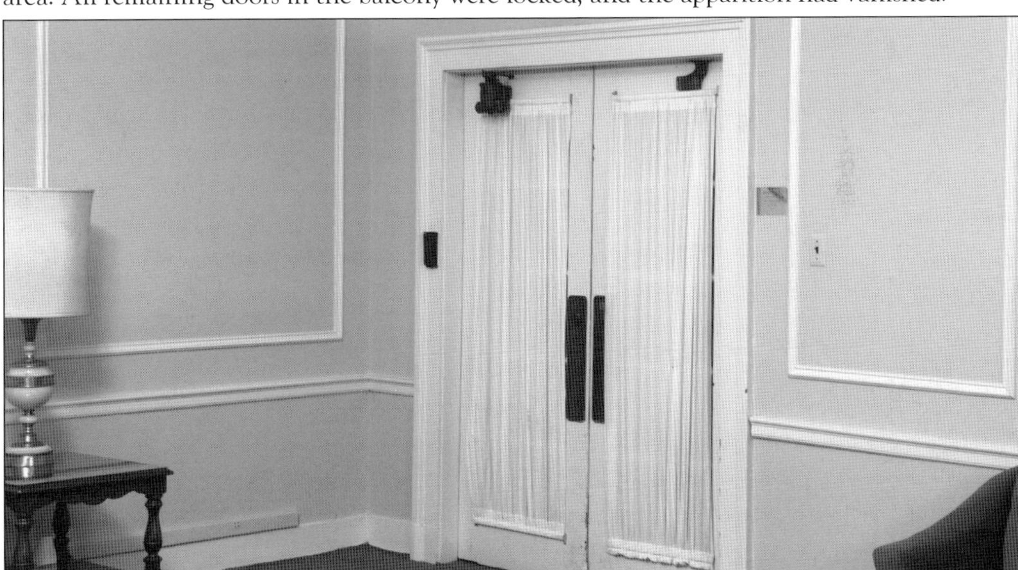

Ghost in Ladies' Lounge. While vacuuming the area, several employees experienced an unexplained shutoff of the lights in the room. Something had turned off the wall switch, but the nearby door had never opened. Many walking through the sitting room also have felt a mysterious chill rattle their bones, adding to the legend of the ghost in the Cathedral. One might ignore this single event, but the presence of multiple sightings adds an aura of validity.

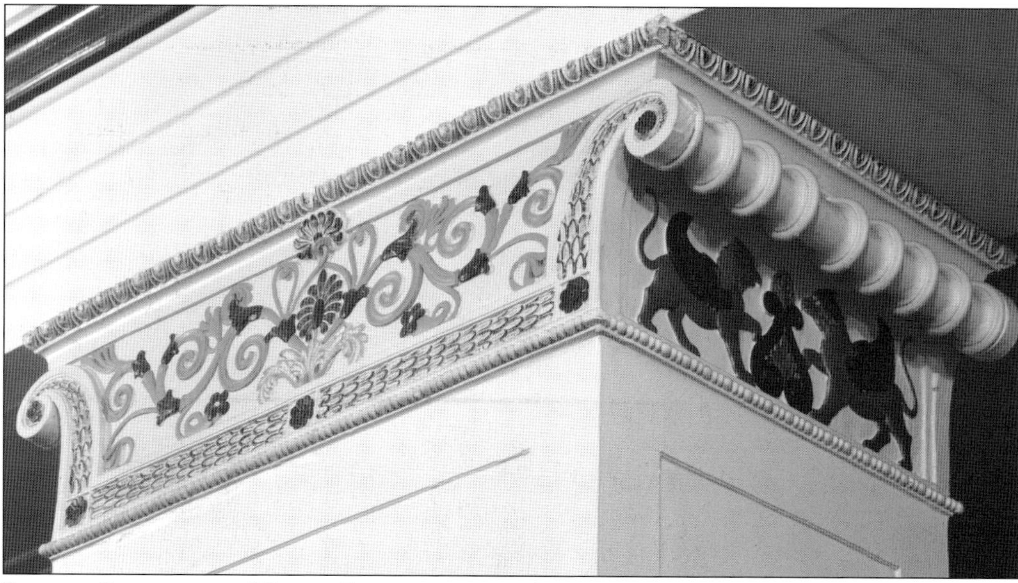

Pillars Capped with Winged Lions. Massive white pillars in the main dining room portray a geometric design and pairs of maroon-colored lions with gold wings. The maroon color repeats itself in the railing background, an upper wall band, and the islands between the steps leading to the Crystal Ballroom.

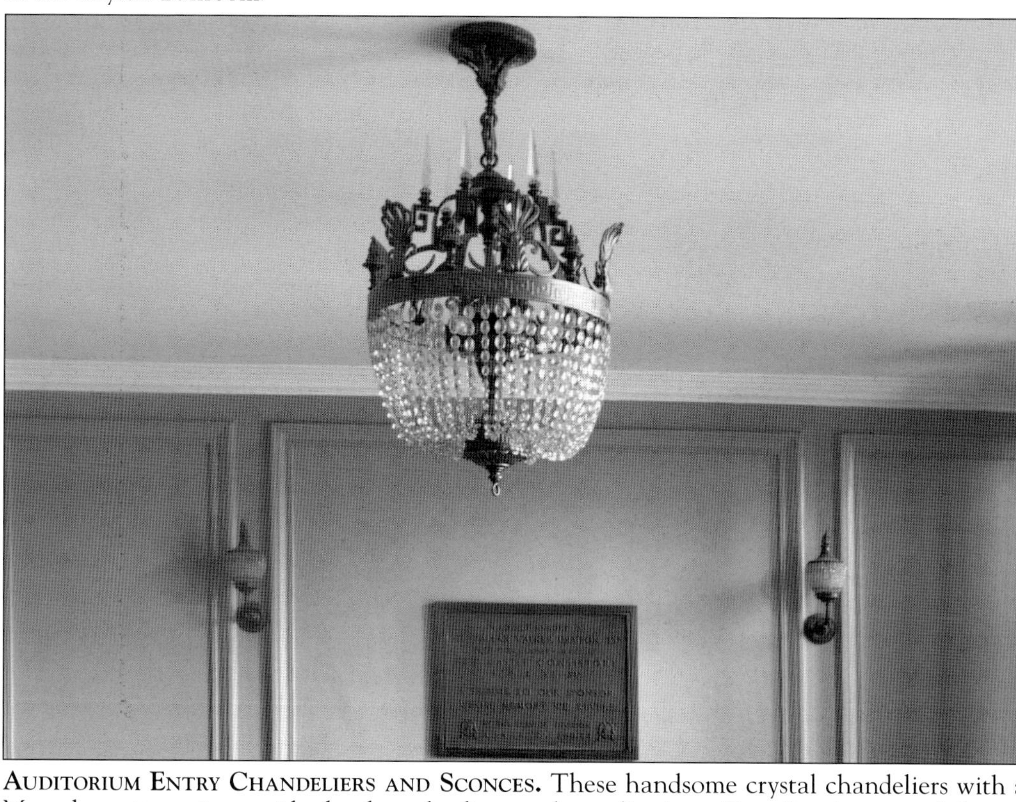

Auditorium Entry Chandeliers and Sconces. These handsome crystal chandeliers with a Meander pattern sits outside the doors leading to the auditorium. Coordinating crystal domed sconces reinforce the dramatic effect of the lighting.

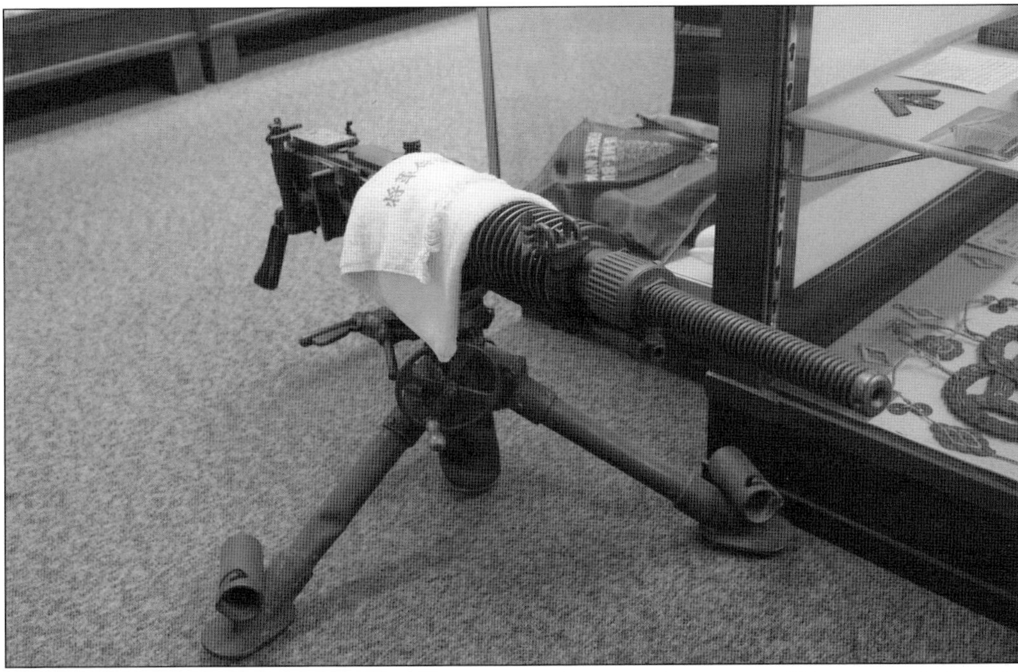

Japanese Machine Gun. The library contains three machine guns. The air-cooled model photographed includes the manufacturer's name in Japanese. Two German machine guns built at the Krupp Works in 1917 and capable of firing 600 shots per minute sit in each corner of the west side of the room.

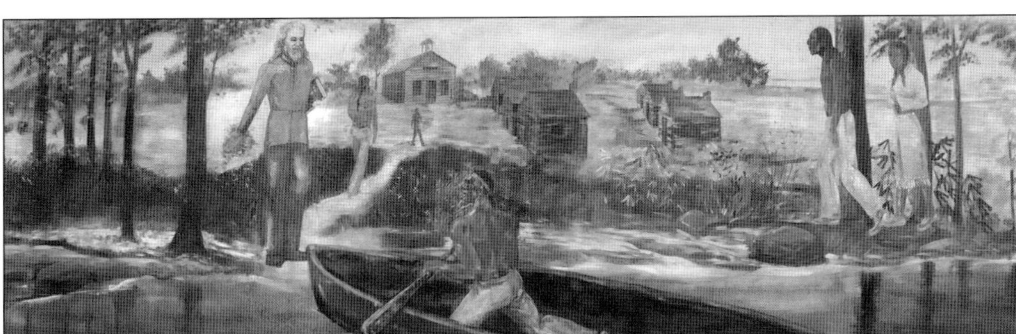

Early Settlers Christianizing Native Americans. A missionary in buckskins carrying a Bible welcomes a Native American in a canoe about to land for an upcoming church service in a small frontier town. Other participants wind their way along a path for the gathering. The Industrial Appraisal Company listed this 20-by-6-foot mural in the west lodge dining area at $20,000. Before recognizing its value, members considered painting over it to brighten up the room.

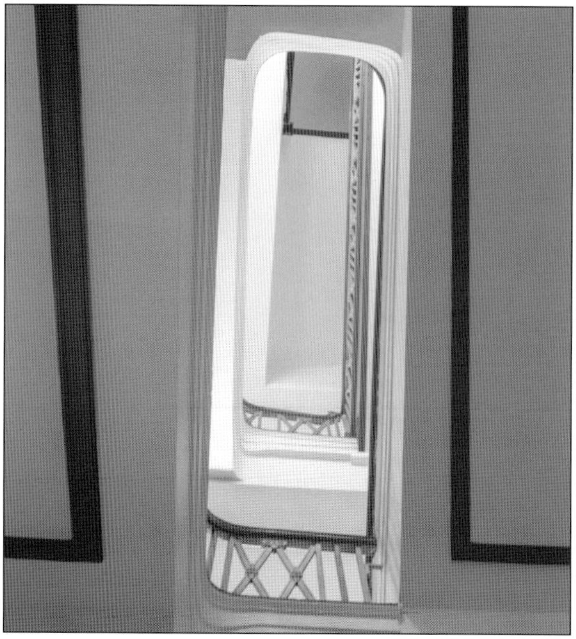

Geometric Vista of Stairway Railings. Photographer Bill Conkle has captured an interesting view of the wrought-iron and wooden railings leading up the stairways of the Cathedral's nine-story structure.

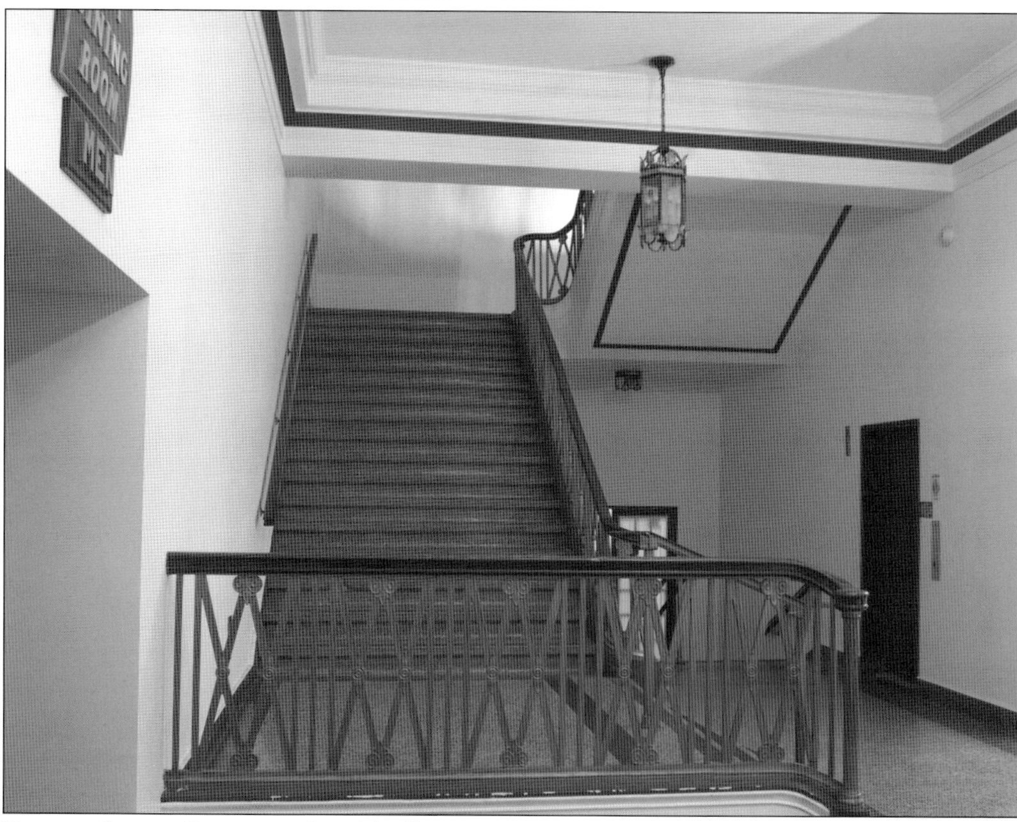

Dining Level Stairway. Marble floors, Masonic symbols, Greek and Roman designs, decorative railings, elaborate chandeliers, and complex doorways make the Cathedral an architectural marvel.

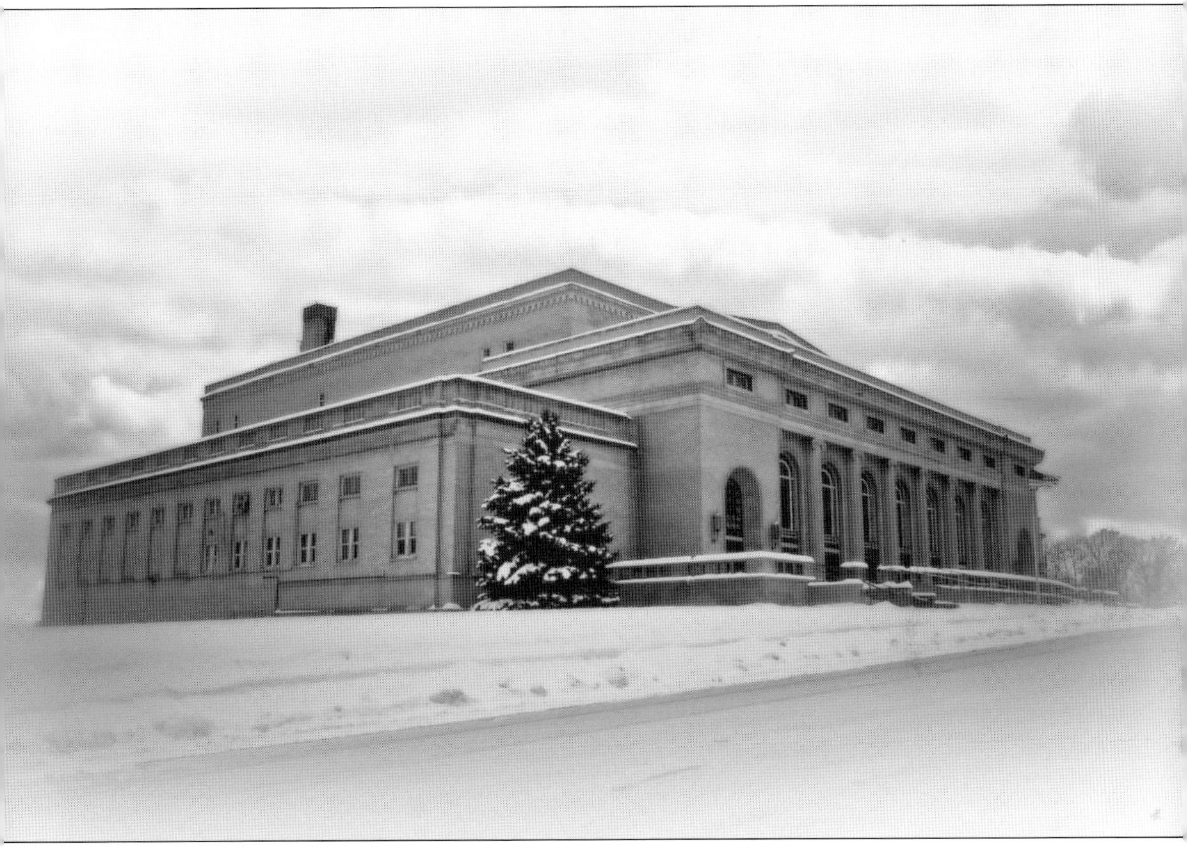

The Scottish Rite Cathedral. This photograph taken on a chilly winter day shows off the Cathedral in all its glory.